DATE DUE

ALSO BY PAT SHIPMAN

The Evolution of Racism

The Wisdom of the Bones
(with Alan Walker)

The Neandertals
(with Erik Trinkaus)

Pat Shipman

TAKING WING

Archaeopteryx and the Evolution of Bird Flight

A TOUCHSTONE BOOK
Published by Simon & Schuster

TOUCHSTONE
Rockefeller Center
1230 Avenue of the Americas
New York, NY 10020

First Touchstone edition 1999
TOUCHSTONE and colophon are registered trade-
marks of Simon & Schuster Inc.

Designed by Sam Potts

Manufactured in the United States of America
1   3   5   7   9   10   8   6   4   2

Library of Congress
Cataloging-in-Publication Data
Shipman, Pat.
Taking wing : Archaeopteryx and the evolution of
bird flight / Pat Shipman.
p.   cm.
Includes bibliographical references and index.
1. Archaeopteryx.   2. Birds—Flight.   I. Title.
QE872.A8S55   1998
568'.22—dc21   97-27527
CIP
ISBN 0-684-81131-6
ISBN 0-684-84965-8 (Pbk)

Acknowledgments

This is not a book I could have written alone, nor would I have had the audacity to attempt it. While no one else bears any responsibility for any errors that may have crept in, many of the accuracies and insights owe much to those who graciously provided reprints or illustrations, gave excellent lectures that I heard, patiently answered E-mail and phone messages, checked parts of the manuscript, and granted interviews. You were all most kind. In particular, I wish to thank Donna Braginetz, Sankar Chatterjee, Luis Chiappe, Philip Currie, John de Vos, Ken Dial, Alan Feduccia, Jacques Gauthier, Jack Horner, Farish Jenkins, Jr., Mimi Koehl, Wann Langston, James Marden, Larry Martin, Diana Matthiesen, Simon Conway Morris, Ulla Norberg, John Ostrom, Kevin Padian, Greg Paul, Siegfried Rietschel, John Ruben, Michael Skrepnick, Sam Tarsitano, David Weishampel, Peter Wellnhofer, and Larry Witmer. If I have omitted anyone from this list, let them be assured it is from poor memory and not ingratitude. Becky Saletan and Denise Roy of Simon & Schuster were enthusiastic, perceptive, and a pleasure to work with, even though this book was not their project from the outset.

On a personal level, I am grateful for the help and support offered by Ginny Armstrong, Nancy Bearsch, Barbara Kennedy, Sally Nevins, my cat Amelia, and my horse Wigston Magna. As always, I owe more than I can ever express to my loving husband, Alan Walker. Thank you all for bearing with me and being there when I needed you.

To Amelia,

whose keen interest
in avian natural history
has often sparked my own.

Contents

TAKING WING

Prologue
A Flight of Fancy

There are seven specimens, and a feather.

It is not much to document the origin of bird flight. Strictly speaking, these specimens are not the entire body of evidence. There are, of course, other fossil birds, not to mention bats, pterodactyls, and insects that have a few things to say about bird flight. And there are living creatures, mathematical models, and aerodynamic theories to help us understand this amazing evolutionary accomplishment. Nonetheless, these seven specimens are crucial. They lie at the heart of complex debates that began with the discovery of the first specimen of *Archaeopteryx* more than 130 years ago and continue up until today. These few, special fossils have served as the basis for brilliant deductions, wild speculations, penetrating analyses, and amazing insights. They have revealed—and continue to reveal—not only the pathway through which birds and bird flight may have originated, but they also tell us much about the strengths and weaknesses of science and scientists.

Only seven precious specimens: as I have learned of them, I have been struck by their paradoxical nature. Seven specimens seem paltry evidence to tell the world everything about a lost animal, yet few extinct species are so well-known. Remarkably, among these seven is perhaps the most beautiful fossil in the world. It is the Berlin *Archaeopteryx,* an exquisite slab and counterslab that capture an extraordinary moment in evolution, when reptiles were turning into birds. No special training is required to see what the Berlin *Archaeopteryx* is; its wings and feathers are as obvious as the teeth in its reptilian jaws and the long, bony tail. The importance of the Berlin specimen cannot be overstated. It is more than a stony record of an extinct species. It is an icon—a holy relic of the past that has become a powerful symbol of the evolutionary process itself. It is the First Bird. Do not think of the specimen as a skeleton in the round, such as the dinosaurs you see mounted in museums. *Archaeopteryx* is more nearly a bas-relief, a two-dimensional sculpture of a fabulous chimera, half-bird, half-reptile. It is a skeleton, plus ligaments, tendons, skin, and feathers, pressed almost flat between the rocks of the ages. Its mode of preservation makes *Archaeopteryx* as much objet d'art as scientific specimen. The difference is that the artist whose metaphorical hand carved this work of art was Time itself.

Time's display is beautifully laid out and daintily detailed. The grinning, toothy skull speaks of death, of birds fused with reptiles, of creatures that no longer exist. The skull is set on a gracefully curved neck, arched strongly backward to show the relentless tightening of tendons in death. On the right and left sides of the slab are delicate feathered wings, evoking a sense of angelic innocence that is disrupted by the three wickedly clawed fingers that adorn each wing. Lower on the slab lie two strong legs with clawed feet, three toes pointing forward on each and a fourth opposed to them, for grasping. To the left lower corner is the impression of the tail, familiar in shape and yet oddly wrong, with symmetrically placed pairs of feathers arising from each vertebra in the peculiarly long tail. While all known specimens of *Archaeopteryx* are remarkable in some way, it is the Berlin specimen that exemplifies them all, combining fragility and grace with enormous emotional and intellectual power.

All known specimens of *Archaeopteryx* come from one special place, in the Bavarian region of Germany. It is a small window through which we can peer at an ancient world (Figure 1). Some 150 million years ago, a few individuals of *Archaeopteryx* died and fell into the still waters of ancient Solnhofen, a shallow lagoon fringed with mud flats and bottomed in fine-grained muds that had been deposited for tens of millions of years. When an *Archaeopteryx* died—of accident, injury, or simple old age—its carcass sank beneath the waters, becoming buried deep in these calcareous sediments, as did the remains of the insects, pterosaurs, fish, crustaceans, nautiloids, and other creatures that lived in or near the lagoon—including a small theropod dinosaur, *Compsognathus*, whose presence proved fateful.

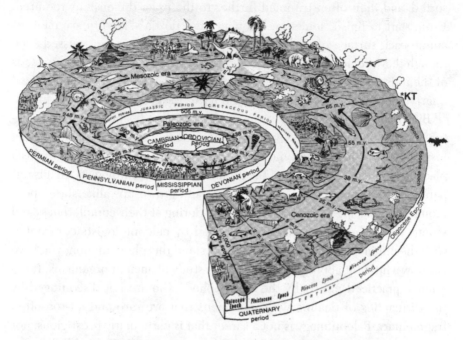

Figure 1. This diagram shows the geological eras and periods. We know that flying vertebrates evolved starting with the Mesozoic era (the Age of Reptiles). The boundary between the Cretaceous and Tertiary periods is marked by a star labeled KT. The impact of a meteor at the K-T boundary may have caused the extinction of all of the nonavian dinosaurs.

The mud sealed it from further decay and destruction, protecting and preserving each anatomical detail as the body was infiltrated and replaced by stone. In time, the sediments consolidated into a fine-grained limestone. Still later, in the nineteenth century, these limestones were prized by humans as the perfect medium for detailed lithographic printing. Solnhofen limestone is so smooth that it will take the sharpest lines, conveying the subtlest textures that can be created by the artist's hand. This happenstance—the value of Solnhofen limestone for printing—was an essential ingredient in the strange story of *Archaeopteryx*. Stone, of course, has many potential uses, and the Solnhofen limestone has been quarried for paving-stones since Roman times. Only in the later nineteenth century did lithography become so important that these wonderfully fine-grained stones were rendered valuable rather than simply useful. As a result, a painstaking process of hand-quarrying (still practiced today) began that led directly to the discovery of *Archaeopteryx* and many other fine fossils. Each slab of Solnhofen limestone is chiseled out by hand, split, inspected for flaws,

15

sorted, and then often trimmed further to the exact dimensions required. From start to finish, sometimes as many as a dozen skilled quarrymen examine each surface of each slab with care, so fossils are not missed even though they are not intentionally sought. Only the coincidence of the needs of the artist and of *Archaeopteryx* accounts for this fact. More mechanized quarrying—the rule elsewhere—would have certainly destroyed the fossils.

But *Archaeopteryx* is more than the world's most beautiful fossil. Its status is singular, despite the seven specimens. It has been a celebrity among fossils almost from the first moment of its discovery. *Archaeopteryx* holds a place in the heart and minds of the public, and of paleontologists, that is unparalleled by any other species. Children and grown-ups alike stare open-mouthed at the specimens on exhibit, wondering at their completeness and strangeness. The same awe is experienced by paleontologists confronted with the original specimens, too. To understand this phenomenon, you have to know a little bit about paleontology, the study of ancient organisms. It is a science practiced mostly by the impassioned, who are not discouraged by the difficulties of dealing with specimens that are rare and all-too-often fragmentary. Paleontology is not a career that is particularly prestigious, nor

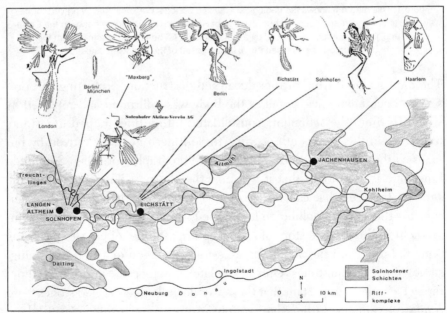

Figure 2. Each of the specimens of *Archaeopteryx* was found in the Solnhofen region of Germany, in Bavaria. Solnhofen was an ancient, shallow lagoon when *Archaeopteryx* was alive, so the geological deposits were formed either by reefs (Riff-komplexe) or sedimentation within the lagoon (Solnhofener Schichten).

16

is it easy; it is certainly not lucrative. But those who choose it can rarely imagine another life that would be as interesting or as much fun.

A paleontologist's fondest hope is to find a complete skeleton of an extinct species in which he or she is interested: an "earliest" or a "best." Careers are made and theories are built on such finds. To find a complete skeleton with an impression of the soft tissues that once surrounded it is a romantic dream that no paleontologist could reasonably hope to fulfill. To have seven such specimens simply defies all probability. Yet that is the case with *Archaeopteryx*. All specimens of *Archaeopteryx* (save the lone feather) are partial skeletons, and a few are nearly complete (Figure 2). Most show feather impressions, for indeed it is the feathers themselves that have been crucial in the history of discovery and recognition. It is the feathers that, from the beginning, have identified *Archaeopteryx* as the earliest bird (although the species has been assigned a number of other taxonomic positions as well). In reality, the chance that one of the very few fossil birds to be so exquisitely preserved was literally *the* first bird is diminishingly small. Nonetheless, deposing *Archaeopteryx* from its throne as First Bird would be a difficult task. Other bird fossils that have been found, and whatever fossils may yet be found, have to vie with *Archaeopteryx* for that title. They will have to rival its completeness and beauty; they will have to be both birdlike and reptilian, without question; they must be undoubtedly ancient as well. It is a daunting challenge. What's more, paleontologists rarely take kindly to the few heretics who have tried to challenge the nearly sacred status of *Archaeopteryx*.

Hovering in the background of any discussion of this species is a scurrilous charge that recurs periodically: that *Archaeopteryx* is a hoax. The accusation, made as long ago as the first discoveries and as recently as the mid-1980s, is that *Archaeopteryx* is not a feathered reptile but a small dinosaur skeleton falsely fitted out with feather impressions. The recent charges have accused putative Victorian forgers, who may have wanted either to prop up evolutionary theory at a time when its acceptance was hotly contested or simply to increase the value of a specimen up for sale. The charge has long been answered and dismissed: the feather impressions are microscopically intricate, showing details of structure of the flight feathers that cannot have been faked in the nineteenth century—and, indeed, would be immensely difficult to fake today. Too, recently discovered specimens have been cleaned and prepared under the light of scientific scrutiny, with no opportunity for illicit modification. Now the forgery charge is rarely spoken of in the literature nor is the entire imbroglio surrounding the charges discussed in scholarly papers or at learned conferences. But the charge of imposture is not forgotten; it casts a pale shadow over the entire topic of *Archaeopteryx* and the origin of bird flight.

The superb preservation of *Archaeopteryx* enabled paleontologists to disprove this dangerous slander. But, ironically, it is also this superb preservation that poses one of the most serious problems to the curators entrusted with these precious skeletons and to the scientists who would analyze them. These specimens simply cannot be treated as any ordinary fossil would be. Preparing the bones fully—that is, freeing them completely from the surrounding rocky matrix—is a standard procedure that would make some aspects of anatomy clearer. But in this case, the bones cannot be fully prepared, for it would destroy the invaluable feather impressions, not to mention the evidence of anatomical articulations and relationships. So the skeletal anatomy of *Archaeopteryx* is tantalizingly complete, yet remains partially veiled. Isolating the bones from one another by removing all of the rocky matrix would destroy the little direct evidence we have of *Archaeopteryx*'s position in life. It would also be tantamount to dismantling a masterpiece, an abomination comparable to cutting the *Mona Lisa* out of its background: unthinkable, too dangerous, too destructive.

When I began to follow the *Archaeopteryx* debates closely, I could not help but muse: how much better could the specimens be? It seemed absurd to me, for my background is in paleoanthropology. Many of the questions in paleoanthropology are similar to those that surround *Archaeopteryx*—How did this form move? What ecological niche was it adapted for? Who were its ancestors?—and similar techniques are often applied. In the course of my research into human origins, I have worked on many collections of fossils pertaining to the record of human evolution and I know from firsthand experience that paleoanthropologists cannot boast of a single fossil as excellent as the Berlin or London *Archaeopteryx*. Despite being in a field often labeled as "contentious," paleoanthropologists can agree on many facets of the anatomy and evolution of the earliest members of our lineage. Thus my background gave me an unusual perspective on the debates over the origins of birds and bird flight. How many more skeletons could paleontologists and ornithologists possibly need before they were able to resolve the basic issues of *Archaeopteryx*'s anatomy and capability? But now I understand the complexities and ambiguities better and I know why *Archaeopteryx* will always remain enigmatic.

The scientists involved with *Archaeopteryx* are not stupid people or inept analysts. To the contrary, they are some of the brightest minds on several continents: keen, dedicated, and clever. But they are handicapped by the very nature of the specimens, by the importance of the specimens' position, and by the singularity of the fossils' preservation. These paleontologists may never be able to see these bones in the round, except perhaps indirectly by sophisticated radiography, because they are too well preserved—and what cannot be seen will always be the focus of argument. The paradox is this: the excellence of preservation makes *Archaeopteryx* especially important by showing the admixture of avian and reptilian features, but that very excellence is also the source of endless frustration and confusion.

And so debate revolves around the positions of the bones in life, how they functioned and how they moved. Did the legs support the body in this position, or were they angled like that? Was the pelvis backwardly tilted, like a bird's, or more forwardly oriented in a more primitive reptilian position? How were the back and neck curved in its habitual posture? Are its feet and toes adapted for perching and grasping tree limbs, or for something else? What are the clawed hands designed for?

Articulation is peculiarly important to understanding the function of avian and reptilian fossils. Where two bones meet each other to make a joint at which movement can occur, the end of each bone has an articular surface. In adult mammals, only a superficial layer of a special cartilage, known as articular cartilage, covers the ends of the bones; the bony tissue itself makes the detailed shape of the joint surfaces. But in birds and reptiles, not only the articular surface but also the entire end of the bone does not ossify and remains cartilaginous. Cartilage, as a soft tissue, does not fossilize in the normal course of events, and so the exact shape of the two surfaces that fit together to form a joint is not preserved in a fossil bird or reptile. Since the shape of those two opposing articular surfaces dictates how the bones that participate in the joint can move, putting the bones of a fossilized bird together into a skeleton and predicting how that skeleton could move are difficult. Without knowing the exact geometry of the joint itself because the soft cartilage has decayed, disarticulated bird or reptile bones can be articulated in a wide range of positions, some of which were surely impossible in life. A fossilized bird skeleton simply provides much less anatomical evidence of that bird's habitual posture than does a skeleton of a fossil mammal. Thus the fact that *Archaeopteryx* is fossilized in a death position (if not in life) is immensely valuable because the bones are unlikely to lie in positions relative to one another that were impossible in life, at least so long as the skin and feathers remain intact. Absurdly, many body parts of *Archaeopteryx* have never been seen in their entirety, even though they are preserved in several specimens. The three-dimensional shape and relationship of some of its bones to others are a constant topic of debate.

Then there are the myriad questions about the abilities of *Archaeopteryx*. Did its wing fold up fully, like a bird's, or only partially? Could *Archaeopteryx* perform a wing flip, a tricky maneuver essential for taking flight from the ground, or not? In fact, could *Archaeopteryx* fly at all? Ironically, there is too much information in the slabs to dismantle them, yet there is too little information to answer these questions definitively.

There is more than this: there is the history of the discovery of *Archaeopteryx*, a tale that cannot be separated from the interpretations of this species. At issue are a series of accidents—literal and intellectual discoveries—that have forever colored the perception of this species. Three specimens of the seven are of special importance: the London *Archaeopteryx*, because it was the first skeleton; the Berlin *Archaeopteryx*, because it is the most complete; and the Haarlem *Archaeopteryx*, because of the extraordinary circumstances and emotional impact of its discovery.

Here is the story of the discovery of a special fossil known as *Archaeopteryx* and the arguments it has inspired.

Chapter 1.

Taking Wing

The very first *Archaeopteryx* to be recognized was a feather impression, dark and clearly delineated on the pale, honey-colored limestone slab. Hermann von Meyer first announced the specimen in 1860. It was a small thing, a feather only about 60 millimeters long and 11 millimeters wide, or about 2.5 inches by one-half inch (Figure 3). This single feather boasted a

central quill or rachis that divided the body of the feather into two asymmetrical vanes, one roughly twice as wide as the other. In this as in much else, the first specimen seemed utterly modern, closely resembling the primary flight feathers of living birds. To be utterly modern 150 million years ago is an impressive feat.

Figure 3. This fossilized feather, found in 1860, was the first specimen of *Archaeopteryx lithographica* to be recognized as an ancestral bird.

No other single, isolated find could have proved so neatly the existence of an ancient bird. Feathered wings are uniquely avian. Flying or flightless, large or small, northern or southern, forest-dwelling, plains-living, or aquatic, all birds have feathered wings and nothing else does. In truth, feathers and wings comprise the quintessence of birdness, even if the point had not been so clearly evident before 1861. There could be no serious question about the sort of creature from which this feather was derived.

Any niggling doubts that there might have been in 1861 were quickly resolved by the finding of one of the three most important specimens, the one destined to be known as the London *Archaeopteryx*. Within a month or two of the finding of the feather, Hermann von Meyer announced the existence of a fossil skeleton of *Archaeopteryx lithographica,* a name he coined that means "ancient wing" and also refers to the lithographic stone that preserves the wing. The suffix "pteryx"—meaning wing—is a common one among the formal names of bird species. By choosing this name, von Meyer did more than immortalize one of the fossil's main features (the feather impressions preserved on the slab). He declared it forever to *be* a bird despite some unmistakably reptilian features. He also made the London fossil the type specimen of a new genus *(Archaeopteryx)* and species *(lithographica).* When a paleontologist believes that he or she has found a new species, he or she formally attaches the name to a particular specimen, the type, which then becomes the exemplar of the features of that species. A type must be formally described in a suitable publication, and that description must both announce the name and offer a diagnosis, an anatomical description that compares the new type with that of other similar species and indicates in which specific ways the new type differs from previously named species. The oldest name given in accordance with the complex and arcane rules of zoological nomenclature has precedence over all others. An individual type specimen carries its name forever, unless someone formally "sinks" that name by demonstrating that the type does not represent a unique species.

But the skeleton of the bird (or not-bird, to those who opposed von Meyer's view) was not in von Meyer's possession. It was part of a collection belonging to Carl Häberlein, a local physician and amateur antiquarian who sometimes accepted fossils in lieu of payment from the quarrymen he treated. In 1862 his hoard numbered 1,703 other fossil specimens, including 23 reptiles, 294 fishes, 1,119 invertebrates, and 145 plants—a fair sampling of the ecosystem in and around the ancient lagoon at Solnhofen so many years before. The catch was that Häberlein had not only a collection but also a beloved daughter, who in 1862 needed a dowry. Häberlein decided to open the question of where *Archaeopteryx* was to nest; he put the entire collection up for sale. As he had doubtless hoped, he provoked a brisk international scrimmage as the prestigious museums of Europe vied to acquire this prize. The feather slab had already gone to the Museum of the Academy of Sciences in Munich, with the counterslab in the Humboldt Museum für Naturkunde in Berlin. Would bird and feather be reunited?

While he waited for offers to come in, Häberlein nurtured an aura of mystery about his *Archaeopteryx* specimen, the crowning glory of his collection. For a while, no one was allowed to make drawings or photographs of it, but this

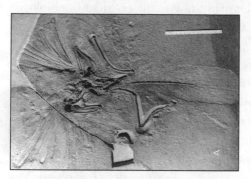

Figure 4. The London *Archaeopteryx*, found in 1861, was sold to the British Museum of Natural History (now the Natural History Museum) for the then-unprecedented price of seven hundred pounds.

secrecy spawned rumors that it was a fake. In a sense, the accusation was hardly surprising, given the half-reptilian, half-avian status the specimen was purported to show. *Archaeopteryx* was a rather neat fulfillment of the predictions of the new and highly controversial theory of evolution first published just a few years earlier. Accidentally finding a perfect transitional form between birds and reptiles was an occurrence too fortuitous and too well-timed to be believed by the skeptical. Eventually Häberlein permitted a select few scholars to inspect his prize, so as to certify its genuineness and heighten interest.

The currency of arguments over evolutionary theory had an enormous effect on the status of Häberlein's *Archaeopteryx*. In England, Charles Darwin's book *The Origin of Species* had been published only two years earlier, in 1859, and florid public debates were raging. The ideas of natural selection and survival of the fittest, forces slowly shaping the hereditary endowments of all living creatures, struck at the very heart of mainstream Christian beliefs in divine creation and the immutability of species. Implicit in this view of the world was a view of the social inequalities of Victorian England, the powerful class system that mirrored the unchangeable hierarchy of the natural world—or so Darwin's opponents thought. Darwin's audacious theory implied not only that humans were mammals, like all others, but that they, too, were subject to the evolutionary laws of Nature.

In 1860, as Thomas Huxley was still stoutly defending his friend Darwin's thesis in England, *The Origin of Species* was published in a German translation made by paleontologist Heinrich Bronn. His was a translation with a difference. Prior to 1860, evolutionary ideas had met with scathing opposition in Germany that was orchestrated by Rudolf Virchow, then the most important man in German biological science. Aware of Virchow's strong disapproval of evolutionary theory, Bronn made two insidious changes in Darwin's thesis. First, he omitted Darwin's most pregnant sentence—"Light will be shed on the origin of man and his history"—and the only one in which he addressed the inflammatory issue of *human* evolution. This omission suggested even to Darwin's few followers (like the biologist Ernst Haeckel) that the Englishman had overlooked the importance of his theory for understanding the human condition and thus diminished the true magnitude of his vision. Second, Bronn appended to Darwin's text a series of his own criticisms of the theory, some fair and others far less so, in an attempt to insure that no intelligent reader would read and accept Darwin's words too readily. Once Bronn's translation was published, Virchow and his followers attacked Darwin's book, as might have been expected. One of their criticisms was that his theory was poorly supported speculation, a curious response to a work notable for its wealth of factual detail. Such was Virchow's influence that his skepticism made it difficult for other scientists to endorse Darwin's work, or even to consider it objectively.

When the first *Archaeopteryx* was found, natural history had been an immensely popular hobby for several decades. Fossils and ferns, birds' eggs and beetles, and other specimens of every sort were being avidly collected and classified by dozens of keen amateurs and many scientists alike. In 1853 and 1854, life-sized models of extinct animals, including a then-new and improved (but still wildly inaccurate) reconstruction of the dinosaur *Iguanodon,* were placed in the gardens of London's Crystal Palace. Originally designed to house the Great Exhibition of 1851, the Crystal Palace was reconstructed at Sydenham, where it and the dinosaurs were a popular tourist attraction. Contemporary accounts describe an elite dinner—invitations were inscribed on the wing bone of a pterodactyl—for twenty-one leading scientists, a dinner that was actually conducted inside one of the iguanodon models. A few years later, in 1859, the publication of Darwin's book catapulted his theory into the intellectual spotlight, and his ideas quickly became the main topic of conversation among the many throughout Europe who had a scientific interest. One of the more telling arguments against *The Origin of Species* was this: if over time there has been a gradual transmutation or evolutionary change of one type of organism to another, *where* are the transitional forms? Indeed, the gap between reptiles and birds was often cited as an unbridgeable gulf for which no intermediate form could be imagined or found. Darwin himself foresaw this objection, entitling the sixth chapter of his book "Difficulties of the Theory" and dealing first of all with the "absence or rarity of transitional varieties," especially in the fossil record.

"But," Darwin writes, voicing many readers' doubts, "as by this theory innumerable transitional forms must have existed, why do we not find them embedded in countless numbers in the crust of the earth?" He answers the question in the tenth chapter, "On the Imperfection of the Geological Record":

> The explanation lies, as I believe, in the extreme imperfection of the geological record. . . . [I]n all parts of the world the piles of sedimentary strata are of wonderful thickness. . . . Professor Ramsay has given me the maximum thickness, from actual measurement in most cases, of the successive formation in *different* parts of Great Britain; and this is the result . . . altogether 72,584 feet; that is, very nearly thirteen and three-quarters British miles. . . . Moreover between each successive formation, we have, in the opinion of most geologists, blank periods of enormous length. . . . Now let us turn to our richest geological museums, and what a paltry display we behold!

The striking contrast between the vastness of the sedimentary record and the scarcity of fossils can only mean this, Darwin reasons: only a tiny proportion of all of the creatures that once lived on the earth have become embedded in sediments, fossilized, and then found and recognized by learned scientists. The expectation of finding a perfectly graded series of transitional forms recorded in the rocks of the earth is thus probably a vain hope. The fossil record preserves only intermittent glimpses of the millions of life forms that have inhabited this earth, not a complete record.

Thus when the first *Archaeopteryx* skeleton was announced to have a mixture of avian and reptilian features, it fulfilled the hope of finding a truly transitional form. The specimen became the cause célèbre of both sides, the believers wanting to display the specimen and prove their point and the nonbelievers hoping to refute its transitional status. Who was to have control of the specimen and thus the immense power of making the first scientific interpretation of the specimen? A fervent attempt was made to keep the specimen in Germany, where it would be placed in the state collection in Munich. But J. Andreas Wagner, a professor of zoology at Munich University and a kindred spirit to the great pathologist Virchow, objected. Wagner could not believe that a transitional form existed between reptiles and birds, and therefore *Archaeopteryx* could not be one. In 1861, although he had not yet seen the specimen for himself, Wagner expressed his views in a paper delivered to the Munich Academy of Science, called "A New Reptile Supposedly Furnished with Bird Feathers." Its title gives away Wagner's main point. He ignored von Meyer's name for the specimen, which had precedence according to the rules of nomenclature, and proposed a new one, *Griphosaurus,* to reflect his view of the specimen as a feathered reptile. Wagner added at the end of his paper:

> In conclusion, I must add a few words to ward off Darwinian misinterpretation of our new Saurian. At first glance of the *Griphosaurus* we might certainly form a notion that we had before us an intermediate creature, engaged in the transition from the Saurian to the bird. DARWIN and his adherents will probably employ the new discovery as an exceedingly welcome occurrence for the justification of their strange views upon the transformation of animals. But in this they will be wrong.

The controversy had taken wing.

Wagner's damning view did little to help the movement to keep *Archaeopteryx* in Germany. In the end, the British Museum (Natural History) procured the skeleton, which was henceforth known as the London *Archaeopteryx.* Together with the rest of Häberlein's collection, the British Museum paid £700, of which an astounding £450 was for *Archaeopteryx* itself. The fee was paid in two installments in 1862 and 1863. This was perhaps the only time in history that a paleontological discovery contributed to a young lady's dowry, and a very handsome contribution it was. A way of evaluating the significance of the £700 paid for the fossils is by comparing the sum to the annual incomes of various figures of that era. In 1862, Thomas Huxley earned £1,124 7s 10d from his writings, lectures, and professorship: on that income, he supported himself, his wife and five children, his widowed sister Ellen, a niece, and paid for finishing school for yet another niece, ending up the year some £227 in debt. If the number of servants employed by the Huxley household was on a scale deemed "suitable" for their income, then they would also have supported a cook, upper housemaid, nursemaid, under housemaid, and a manservant. (In Victorian England, the number of servants was calibrated to income and status. For example, a "wealthy nobleman" would appropriately employ twenty-six or more servants in various categories, and even modest households, with incomes of £150 or £200 a year, would employ a "maid-of-all-work" at £9 to £14 per annum and a girl "occasionally.") Thus the price paid for the total Häberlein collection was about one-half of what Huxley and his rather large extended family needed to live in reasonable middle-class comfort. Moving down in social scale a few notches, the Huxleys would appropriately have paid their cook—one of the highest ranked and paid female servants—about £14 to £40 a year, and slightly less if an extra allowance were made for her consumption of "tea, sugar, and beer." Moving up the social scale, Charles Darwin can serve as an example. Darwin came from a far wealthier family than Huxley's, married a daughter of the monied Wedgwood family, and always had a substantial income from his writings and investments; in 1861, these provided about £8,000, an amount roughly eleven times the price of the Häberlein fossils, seven times the amount of Huxley's annual income, and two hundred times the annual wages of a well-paid cook.

The keeper of geology at the British Museum who had advocated the extravagant purchase of the specimen was Robert G. Waterhouse. Under the circumstances, Waterhouse might reasonably have been expected to study the specimen himself, but he was interested primarily in entomology and not at all in fossil birds. Thus the privilege of describing and publishing about the specimen fell to Richard Owen, a prominent anatomist and paleontologist who was then superintendent of the Natural History Departments at the British Museum.

Owen was a lean-faced, pop-eyed man of pedantic turn of phrase; he was also notorious for using ridicule and malicious attacks to further his own position. Yet he could rightfully lay claim to many accomplishments, among which were supervising the construction of the cement dinosaur models at the Crystal Palace at Sydenham, predicting the shape of the femur or thigh bone of the giant extinct moa (then new to science) from a fragment, publishing numerous anatomical and paleontological papers, and serving on innumerable royal commissions and committees designed to encourage Science with a capital S. As an intimate of the royal family and a conspicuous public figure, Owen was a dangerous man to cross. He was not, however, a popular one due to his high opinion of himself, which contrasted sharply with his low opinion of everyone else, and his avidity to steal credit for someone else's work whenever possible. Hugh Falconer, a fellow British paleontologist, spoke for many when he described Owen as "ambitious, very envious and arrogant, but untruthful and dishonest." Understandably, Owen had few friends in London scientific circles, although many acknowledged his formidable intellect. Even the mild-mannered Charles Darwin had harsh things to say of Owen in his *Autobiography*, though the rise in Darwin's theory had much to do with the decline in Owen's power in scientific circles. Owen was a man always ready to pronounce judgment on any topic, whether it fell within his expertise or not. But on the subject of evolution, which surely did fall within his field, Owen seemed generally opposed. However, he repeatedly contradicted himself on this matter, expressing his muddled thoughts in prose that was often dull and sometimes simply incomprehensible. In contrast, Darwin's defender Thomas Huxley was witty, intelligible, and a gifted public speaker. Huxley's irreverence, combined with his delight in a good argument, caused him to take a special pleasure in skewering one so pompous as Owen, whom he frequently bested in written and verbal debate.

Archaeopteryx provided another opportunity for a spirited skirmish between the two. By 1863, Owen had completed a monographic study of this new and most important specimen. He insisted on giving it yet another scientific name—*Archaeopteryx macrura,* or long-legged *Archaeopteryx*—although there was nothing wrong with von Meyer's *Archaeopteryx lithographica.* Owen declared the specimen to be "unequivocally a bird" with some characters seen only in the embryos of modern birds and yet other features indicating "a closer adhesion to the general vertebrate type." The London *Archaeopteryx* is neither as complete nor as neatly articulated as the Berlin specimen, which would not be found until 1877, but it was by any paleontologist's standards a wonderful fossil. And, as Wagner had predicted in his scathing dismissal of *Archaeopteryx,* Darwin's adherents—especially Thomas Huxley—were anxious to see if it might be used as a transitional form to illustrate their hypothesis while Darwin's opponents, like Owen, found no such suggestion in the fossil. Huxley soon embarked on a thorough study of the specimen himself, notwithstanding Owen's 1863 publication, and by 1868 was ready to give his own masterful exposition of the specimen. As ever, it brought him into open conflict with his old enemy, Owen.

Huxley's challenge is clear from his very title, "Remarks upon *Archaeopteryx lithographica*," for he shuns Owen's new name. His opening remarks are a model of circumspection:

The unique specimen of *Archaeopteryx lithographica* (von Meyer) which at present adorns the collection of fossils in the British Museum, is undoubtedly one of the most interesting relics of the extinct fauna of long-past ages; and the correct interpretation of the fossil is of proportional importance. Hence I do not hesitate to trouble the Royal Society with the following remarks, which are, in part, intended to rectify certain errors which appear to me to be contained in the description of the fossil in the Philosophical Transactions for 1863.

At this point, a footnote identifies the description as having been written by Owen, a fact that was surely known to all of the Royal Society members who read the publication or were in attendance when Huxley read the paper aloud. Huxley continues in a tone of sweet reasonableness:

It is obviously impossible to compare the bones of one animal satisfactorily with those of another, unless it is clearly settled that such is the dorsal and such the ventral aspect of the vertebra, and that such a bone of the limb-arches, or limbs, belongs to the left, and such another to the right side.

Identical animals may seem to be quite different, if the bones of the same limbs are compared under the impression that they belong to opposite sides, and very different bones may appear to be similar, if those of the opposite sides are placed in juxtaposition.

This preamble sets out some very basic principles of anatomical analysis and comparison, points with which no one would argue. But the preamble is well-planned, for Huxley proceeds to demonstrate that Owen has misoriented the vertebral column, mistaken the bones of the left leg and foot for the right, taken the left portion of the pelvis for the right, misidentified the right scapula (or shoulder girdle) and humerus (or upper wing bone) as the left, and has incorrectly oriented probably the most important bone in the specimen, the furcula, or wishbone. (The mammalian equivalent to a furcula would be a fused pair of clavicles, or collarbones.) Owen's is a horrifying series of errors; very little in his interpretation is correct. Huxley's remarks make Owen look either foolish or appallingly sloppy in his work—and no wonder that Huxley rejected Owen's nomenclature, if such was the standard of the latter's scholarship!

29

At the end of the paper, Huxley makes two prescient comments addressing issues that soon became important. First, Huxley challenges one of Owen's conclusions about the skull of *Archaeopteryx*. Because *Archaeopteryx* demonstrably has feathers, Owen inferred that it would have had a toothless (and lipless) beak for preening. After Owen's publication, but before Huxley's, Sir John Evans wrote an article describing a fragment of upper jaw, preserving four teeth, found on the same *Archaeopteryx* slab. Some skeptics rejected the idea that the jaw was in fact part of the skull of *Archaeopteryx* and favored its being a fragment of another creature, for birds do not have teeth. While Huxley does not discuss Evans's jaw directly, perhaps because the state of disarticulation of the London specimen makes its association with the rest of the skeleton questionable, he challenges the solidity of the link between feathers and toothless beaks. Huxley offers examples of various combinations of skull and integument features in extant animals, some of which make it plausible that *Archaeopteryx* may have had teeth. For example,

> The soft tortoises (*Trionyx*) have fleshy lips as well as horny beak; the *Chelonia* [turtles] in general have horny beaks, though they possess no feathers to preen; and *Rhamphorhyncus* [a pterodactyl] combined both beak and teeth, though it was equally devoid of feathers. If, when the head of *Archaeopteryx* is discovered, its jaws contain teeth, it will not the more, to my mind, cease to be a bird, than turtles cease to be reptiles because they have beaks.

Indeed, when the next, more complete specimen of *Archaeopteryx* turned up, in 1877, Huxley and Evans were duly proven right on the matter of toothed jaws.

Huxley's final point addresses the evolutionary relationships of *Archaeopteryx*. All birds possess a foot and pelvis like *Archaeopteryx* as well as feathers, while no reptile shows all of these features. But the pelvis of *Archaeopteryx* resembles that of several known dinosaurs, and the foot, Huxley observes, is surely very similar to the reptile *Compsognathus,* a small dinosaur also known from the Solnhofen limestone. The Solnhofen *Compsognathus* skeleton, preserved in a pose strikingly similar to that of *Archaeopteryx,* is a species and specimen that became critical to the debates in the ensuing decades.

Later in 1868, Huxley reiterated his point with a paper entitled "On the Animals which are Most Nearly Intermediate Between the Dinosaurian Reptiles and Birds." In it, Huxley enumerates the differences between birds and reptiles. Within birds, the great division is between the ratites— the large, nonflying birds like the emu, ostrich, and rhea—and the carinates, all flying birds. Although the former are clearly closer to reptiles, there remains a gap between the two birds and reptiles, a gulf that separates these two great classes of animals and cannot be filled by any living forms. The problem can only be resolved by recourse to the fossil record, Huxley deduces, of which two questions must be asked.

1. Are any fossil birds more reptilian than those now living?
2. Are any fossil reptiles more bird-like than living Reptiles?

And with, as one Huxley scholar has called it, "an air of pulling rabbits out of a hat," Huxley answers both of these questions in the affirmative. As Huxley then shows, *Archaeopteryx* is the most reptilian bird, the link between ratites and dinosaurs, and dinosaurs are the most birdlike of reptiles. All that is needed is an "abracadabra" to complete his performance.

Deriving birds from dinosaurs was a brilliant idea and he was not yet done with it. In 1870, Huxley produced another paper, "Further Evidence of the Affinity between the Dinosaurian Reptiles and Birds." The occasion is his study of a fossil skeleton of the dinosaur *Megalosaurus* and his correspondence on that species with Professor John Phillips at Oxford. Between the two of them, they straighten out some misidentifications among the bones of *Megalosaurus,* as Huxley has already done with *Archaeopteryx.* In this case, portions of the pelvis had been mistaken for parts of the shoulder girdle (but not by Owen) and the humerus was not recognized as such due in part to the fact that it was much smaller than the femur. Once the bones are correctly identified, Huxley is then able to compile an impressive list of anatomical features in which *Megalosaurus* resembles an ostrich-like bird more than other reptiles, such as crocodiles or lizards.

One of the fundamental principles of comparative anatomy is that animals with similar functions will show anatomical resemblances; their "equipment" will be similar if the "task" they are performing is similar. Thus, for example, birds, bats, and pterodactyls all have wings because they fly. But as evolutionary theory began to enlighten comparative anatomy, a second great principle was realized: the degree of anatomical similarity may reflect evolutionary relationships as well as function. Two closely related species that share similar habits will show great and detailed similarities in the anatomical adaptations to those habits. For example, all birds are relatively closely related to each other and all have structurally identical wings, although they may differ slightly in size and shape. But if two species have similar habits yet are only distantly related, then they may show functionally equivalent anatomy with telling differences in the details that reveal the remoteness of the relationship. Bats, birds, and pterodactyls all have wings supported by the elongated bones of their arms and hands, but each has its own unique structural features. Bats, for example, have short arms and five greatly elongated fingers on each hand. Between each pair of fingers is stretched a largely naked skin wing that is continuous with a membrane stretching from the arm to the hindlimb. In contrast, pterodactyls have arms of moderate length and a skin wing that stretches from the body, along the edge of the arm, and out along a single, exceptionally long fourth finger. The pterodactyl hindlimb is probably not uninvolved in anchoring the skin wing, although this idea, like so much else, is debated. While the fourth finger supports the wing, the other fingers of the pterodactyl, much shortened, form grasping claws located near the wrist joint. Although bats and pterodactyls might both fairly be called skin wings, the details of how those wings are made, anchored, and used are different. For their part, birds have shortened and fused the three remaining fingers of each wing (two have been lost evolutionarily) and have covered their wings in feathers. Once again, birds have evolved a unique set of flying adaptations different from those of bats or pterodactyls. These differences demonstrate an important evolutionary convergence; the three groups have evolved in different ways to perfect a similar function, flying.

In other words, the degree and complexity of anatomical similarities allows us to evaluate whether two species are showing relatively superficial resemblances—convergences—or deeper ones that attest to a recent common descent. Evolutionary schemes or phylogenies are formally based on lists of traits that enumerate the detailed anatomical resemblances among a group of species. Most modern paleontologists follow a rigorous set of procedures developed in the 1950s (and popularized in English translation in the 1960s) by German biologist Willi Hennig. In Hennig's approach, all of the observable anatomical features of each of several species are enumerated; those that share traits in common are then clustered, commonly now by a computer program. Eventually, dichotomies develop between primitive ancestral features (held in common by all members of a group sharing a common ancestry) and subclusters comprising a set of shared, derived, or more specialized features. Derived features are also sometimes known as evolutionary novelties, meaning that they are new developments added onto the ancestral structures. By clustering those species that share specific resemblances in a hierarchical fashion, scientists can use these dichotomies to build a branching tree called a cladogram; Hennig's technique is usually called cladistics. A cladogram may parallel and is hoped to closely resemble the true evolutionary tree or phylogeny that expresses the historical sequence of branching events during evolution. Indeed, in an ideal world, any cladogram would be identical to the true phylogeny, but such a circumstance assumes a complete and accurately interpreted fossil record as well as the addition of chronological information that is currently excluded from cladograms.

Even before Hennigian taxonomy became popular, the basic principles of deducing phylogenetic relationships were similar. The more anatomical features, or *characters* in Hennigian terminology, two species have in common, the more closely they are presumed to be related. Polarity of those resemblances is also important. Characters may be primitive—that is, believed to be ancestral—or derived, meaning that they represent a later specialization. A wide group of lineages will share many primitive characters, whereas an ever narrower group will hold derived features or evolutionary novelties in common as evolution proceeds. In the end, each individual species will have a completely unique set of derived features, as well as many retained primitive ones. Deducing the correct phylogeny for a group of species thus involves an exercise in logic and careful comparative anatomy.

The anatomical details or characters cited by a paleontologist or anatomist constitute the evidence, which ultimately adds up to a certainty approaching proof. Thus, for example, although birds like the penguin or the ostrich are incapable of flight—the archetypal specialization of birds—they share so many characters with flighted birds that their close relationship cannot be doubted. Abandoning flight and adapting to a terrestrial (for the ostrich) or semi-aquatic (for the penguin) way of life have led to the evolutionary replacement of some of those primitive avian features with derived ones peculiar to the ostrich and penguin lineages respectively.

Huxley used a similar overall logic in deducing relationships. He pointed to many close resemblances in the hindlimb "equipment" of birds and dinosaurs. For a start, the disproportion of fore- and hindlimbs suggests *Megalosaurus* to have been predominantly if not exclusively bipedal, in a stance that matches presumed dinosaur footprints discovered in the Weald and is, of course, practiced by all birds. He cites details of the pelvis, the femur or thigh bone, the two bones of the second segment of the leg (the larger tibia and the more slender fibula), and the foot.

A quick comparison of avian anatomy to human anatomy (Figure 5) will make Huxley's points clearer. Humans are more generalized (or less specialized, depending on your point of view) vertebrates than birds or dinosaurs. We have seven ankle or tarsal bones that articulate proximally (at the top end) with the bones of the lower leg and distally (at the bottom end) with the bones of the foot proper. The two largest tarsals are the astragalus, which permits the foot to pivot up and down at the ankle, and the calcaneum, or bony heel. The other five tarsals (the navicular, the medial cuneiform, the intermediate cuneiform, the lateral cuneiform, and the cuboid) form the beginning of the arch of our foot. These smaller tarsal bones articulate in turn with five metatarsals, the elongated bones that

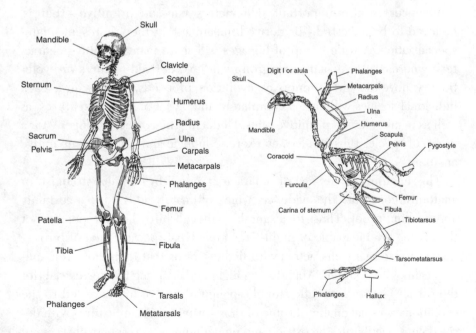

Figure 5. Compared to the skeleton of a generalized vertebrate like a human (left), a bird's skeleton (right) has been modified because of the demands of flight. In particular, the bones of the bird's wrist, hand, ankle, and foot have been reduced in number and altered in shape.

make up most of the arch of our foot. Distally, each metatarsal then ends in a series of phalanges that make up our bony toes.

Birds have more specialized legs and feet than ours, with a reduced number of bones. The femur is short and tucked up close to the body; the femur and its muscles form the bird's thigh. The tibia and fibula (reduced to a slender splint bone) make up the segment known in chickens or turkeys as the drumstick. In an adult bird, the drumstick or tibiotarsus includes the astragalus and calcaneum that have fused to the end of the tibia, but these are separate and largely cartilaginous bones in young birds like those found in the grocery store. The smaller tarsals and the metatarsals have fused together to form a single, elongated bone known as the tarsometatarsus, which makes up the third segment of the bird's leg leading to the foot. This segment, covered in scales, is generally removed in prepackaged grocery store birds but can be seen on any inhabitant of a backyard bird feeder. The phalanges are usually reduced from five to three or four digits; the bird actually stands on its toes, not on its entire foot. Most dinosaurs follow a birdlike pattern of leg and foot anatomy rather than a mammalian one.

Huxley knew bird anatomy well, having only a few years before written a major monograph on birds. In birds, there is a strong bony crest that runs from top to bottom, down the anterior surface of the tibia; there is another bony ridge running down the lateral side where the fibula attaches. Huxley observed both of these ridges in the tibia of the dinosaur *Megalosaurus*. The distal or ankle end of the tibia articulates with the astragalus, and a special bony flange of the astragalus, technically known as the ascending process of the astragalus, sticks up and fits neatly into a concavity on the front of the tibia. This ascending process appears to buttress the front of the tibia. Huxley illustrates the anatomy of a young ostrich, in which the astragalus has not yet fused to the tibia (though the two are fused in adults), and compares it with the same bones in *Megalosaurus*. They are remarkably similar in anatomy, although the ascending process of the astragalus in this dinosaur apparently never fuses to the tibia. In other, smaller species of dinosaur, like *Compsognathus* or *Ornithotarsus* (or bird foot), the two bones do eventually fuse into a birdlike condition.

To emphasize the extremely close resemblances between avian and dinosaurian anatomy, Huxley resorts to a humble but compelling comparison.

> I find [he writes] that the tibia and astragalus of a Dorking fowl
> remain readily separable at the time at which these birds are

usually brought to table. The cnemial epiphysis [the crest on the tibia] is also easily detached at this time. If the tibia without that [crest] and the astragalus were found in a fossil state, I know not by what test they could be distinguished from the bones of a Dinosaurian. And if the whole hind quarters, from the ilium [of the pelvis] to the toes, of a half-hatched chick could be suddenly enlarged, ossified, and fossilised as they are, they would furnish us with the last step of the transition between Birds and Reptiles; for there would be nothing in their characters to prevent us from referring them to the *Dinosauria.*

These were prophetic words, as events just more than one hundred years later would show.

By the time the next specimen, the Berlin *Archaeopteryx,* came to light, the outlines of the debate had been sketched. *Archaeopteryx* was, or was not, the proof of evolutionary theory. It was, or was not, either the first bird, or a bird-reptile chimera, or a feathered reptile. It was, or was not, so like a reptile that an unfeathered specimen might be taken for a small bipedal dinosaur, such as *Compsognathus.*

The gorgeous Berlin specimen of *Archaeopteryx* was also found at Solnhofen, in 1877 (Figure 6). Within weeks, it came into the hands of Ernst Häberlein, the son of Dr. Carl Häberlein, who had sold the London *Archaeopteryx* to provide his daughter with a dowry. Ernst Häberlein bought the new specimen from the quarry manager for a paltry 140 deutsche marks intending, as his father had done, to sell *Archaeopteryx* for a small fortune. It was a highly mar-

Figure 6. The Berlin specimen of *Archaeopteryx*, found in 1877, is the most complete and the most beautiful.

ketable commodity, as the second known skeletal specimen of *Archaeopteryx* (and, after all, the feather might not actually have come from *Archaeopteryx* but from some other feathered species). By now, *Archaeopteryx* was the subject of many controversies and much scientific attention, and no other feathered species would ever surpass its importance. Since the new find was an even more wonderful and complete specimen than the London *Archaeopteryx,* Ernst Häberlein had little difficulty in attracting potential buyers.

One of the first offers came from America. Working through a friend in Dresden, paleontologist O. C. Marsh made an offer to acquire the specimen for the Yale Peabody Museum for 1,000 deutsche marks, a substantial sum in those days. Marsh's bid was refused, presumably because Häberlein thought he could do better than this offer, which was an almost tenfold profit over his initial outlay. Despite the keen interest in the specimen, however, buyers with hard cash proved hard to come by. Subsequently, in a letter dated March 7, 1879, F. A. Schwartz of Nürnberg (Nuremberg), acting for Häberlein, made a counteroffer to Yale suggesting the much more magnificent price of $10,000—the equivalent today of several million dollars, as judged by a modern-day paleontologist. Schwartz enclosed a beautiful tracing of *Archaeopteryx* with the letter, one of the first representations of the new specimen.

There is no indication in the Yale archives why Marsh did not negotiate further. It would have been satisfyingly fitting if Yale University had acquired the treasure, since a later professor at Yale figures so prominently in recent work on *Archaeopteryx.* Marsh was something of a penny-pincher, and perhaps $10,000 seemed too high a price, even for such an excellent specimen. Perhaps he was also influenced by a newspaper clipping from Nürnberg, preserved in the Yale archives along with Schwartz's letter, in which a Professor Giebel in Halle remarks that the specimen may be a fraud. Whatever Marsh's reasoning, in the end, the specimen was bought by Werner Siemens, a German industrial magnate, who in turn sold it to the Prussian Ministry of Culture for 20,000 deutsche marks. This was an amount substantially lower than the $10,000 Schwartz wanted Marsh to pay, but much higher than the 1,000 deutsche marks Marsh had originally offered.

Thus the Berlin *Archaeopteryx,* as it was henceforth known, found a permanent roost in the Humboldt Museum für Naturkunde. Despite the hefty price tag, the Berlin *Archaeopteryx* was not formally described until 1897 (almost twenty years later) by Wilhelm Dames. As by now seemed to be standard procedure, Dames coined yet another name, *Archaeopteryx siemensii*. This was more than a tactful attempt to honor the patron who facilitated its purchase. The new name reflected Dames's belief that the specimen was similar enough to the London *Archaeopteryx* to belong in the same genus but different enough to be classified as its own species. This point of view has not been well-supported over the years, and nearly all scientists today would place the specimen in *Archaeopteryx lithographica*.

After this initial flurry of finds, no new specimens of feathers or skeletons of *Archaeopteryx* were discovered until 1955. In that year, another partial skeleton was found at Solnhofen. This one was badly decayed and disarticulated prior to fossilization; it is considerably more fragmentary and has only faintly visible feather impressions. This find became known as the Maxberg specimen, since it was exhibited in a museum of that name near Solnhofen for some years (Figure 7). It has since been removed from public exhibition by its owner, Eduard Opitsch of Pappenheim, and the Maxberg specimen is no longer available for scientific study.

Figure 7. The Maxberg *Archaeopteryx*, found in 1955, includes only the torso and some feathers of a badly decomposed individual.

But in 1970, an amazing thing happened: the third influential discovery of *Archaeopteryx* was made. The right man, standing in the right place—in fact, perhaps the *only* place where such a discovery might be made—found another partial skeleton of *Archaeopteryx* that had been hiding for years. The man was John Ostrom, the second Yale paleontologist to be linked to *Archaeopteryx*, and the place was the Teyler Museum in Haarlem, Holland. The hiding place was a glass case on public display, complete with a handwritten label. But its camouflage was near-perfect, for its label identified the specimen as part of a pterodactyl.

Ostrom had long been fascinated by the evolutionary development of flight. Flying seems such an impossibly difficult task, requiring controlled, powered movement through the air—a distinctly unforgiving medium. Humans have yearned to fly for centuries—some of our most ancient myths, like that of Icarus, speak of the metaphorical and literal dangers of flying too high. It is only in the last century that we have achieved powered flight and been able to conquer the air by means of sophisticated technologies that took many years of trial-and-error experimentation, often ending in spectacular crashes. More than one famous would-be aviator died in his attempt to devise a machine capable of flying, and even more died in the first few decades after flight had been accomplished but was being perfected. Flying is a precarious mode of locomotion and one that exacts a painful price for miscalculation or failure: so the history of human endeavors to fly has taught us.

Ostrom knew this, but he also knew that not only birds but also bats and pterodactyls evolved flight independently. This evolutionary experiment, so difficult for humans with all their intelligence and all their technology, was successful not once but three separate times. It is one of the most stunning adaptations evolved by any creature anywhere, and John Ostrom wanted to know *why* and *how*.

At Yale, Ostrom's interest was sparked by a large and fine collection of pterodactyls (but, alas, no *Archaeopteryx*) that had not yet been analyzed functionally in order to determine how they flew. Ostrom proposed to do just that. He needed to compare his Yale pterodactyls with others around the world, and he went to Haarlem in 1970 to see the specimens there. He was striving to understand the living machinery with which pterodactyls propelled themselves through the air. Did they all fly in the same manner? What features varied, which were constant? He was a man with a burning question and an open mind.

His quest led him to the Teyler Museum. This special setting played a significant role in Ostrom's recognition of the concealed *Archaeopteryx*. Today, John de Vos, the present curator of paleontology at the Teyler, calls it "the most beautiful museum in Europe" with justifiable pride. A handsome stone building overlooking the River Spaarne, the museum was once the private home of Pieter Teyler van der Hulst, a wealthy Dutch industrialist who died in 1778 at the age of seventy-six. He bequeathed his fortune to a foundation to further the progress of the arts and sciences—and, not incidentally, to surpass the efforts of a scientific society that met in a house across the river and had rejected him for membership. The museum opened in 1784 and is the oldest public museum in the Netherlands, as well as the most beautiful. Each room is preserved as it originally was, from the lovely auditorium that holds 150 listeners on glistening, horsehair-covered seats to the oldest part of the museum, the striking Oval Hall topped with an enormous domed window. The library houses an incomparable collection of rare natural history books and periodicals on beautiful wooden bookcases and a balcony decorated with wrought iron; the art collection features exquisite Rembrandts and some of Michelangelo's studies for the Sistine Chapel; the physics laboratory houses a superb collection of machines, electrostatic generators, engines, and the like; the coin and medal collections are the second most important in the Netherlands. Another of the Teyler's renowned treasures, the one that drew Ostrom, is a priceless collection of fossils and minerals. There are fossil cave bears and saber-toothed tigers, whalelike mosasaurs, the first "fossil human" ever described (actually a giant salamander known as *Homo diluvii*, or "the man of the Flood"), and pterodactyls, among other wonders. The fossil exhibits are all maintained in their original cases, with the original hand-lettered labels from the early years of the twentieth century; these labels were an innovation since many natural history exhibits were still simply jumbled collections of oddities and curiosities. The Teyler's specimens are still arranged in their original deliberate groupings that reflect either the animal preserved or its place of origin.

In keeping with the philosophy of being a "museum of a museum"—one that preserves the architecture, furnishings, and exhibit styles of the past— the Teyler is a daylight museum. Thus the public areas are lit only by sunlight that streams in through the large windows (although the administrative offices have electricity, and modern plumbing has been installed). Gaslights or candles, once an auxiliary source of illumination, are now considered too dangerous; an experiment in issuing visitors flashlights provoked an enormous expense, as the flashlights too often departed with the visitors. Today, no other source of light supplements the sun. On dark winter days, the museum simply closes early. The impression is charming and effective as a way of displaying the Teyler's treasures. It also proved to be crucial to the discovery of one of the Teyler's most precious fossil specimens, its *Archaeopteryx*.

Ostrom had come to study pterodactyls, of which the museum had a number of specimens including the type of a species called *Pterodactylus crassipes*. Hermann von Meyer—the same man who found the feather and named the London specimen *Archaeopteryx lithographica*—proposed this name in 1857 for a slab and counterslab found in the Solnhofen limestone in 1855. For Ostrom's purposes, inspecting this particular specimen of *Pterodactylus crassipes* was crucial if he wanted to be able to recognize other specimens of this species that were either as-yet unidentified or incorrectly identified, for the Teyler *Pterodactylus crassipes* is the type specimen. Since *P. crassipes* was considered in 1970 to be a perfectly valid species, Ostrom had to examine it closely for his study.

On that day in August 1970, Ostrom's host was the curator C. O. van Regteren Altena. Van Regteren Altena brought Ostrom the slab and counterslab and set him up with a chair and a table to work on. The curator was then called away to attend to some work of his own, leaving Ostrom to familiarize himself with the specimens; van Regteren Altena courteously promised to return in a few minutes to see if Ostrom had everything he needed. As soon as Ostrom picked up the slab, he knew immediately that the anatomy was all wrong for a pterodactyl. He had been studying pterosaurs, the general group into which pterodactyls fit, for months, and knew them well; this wasn't one.

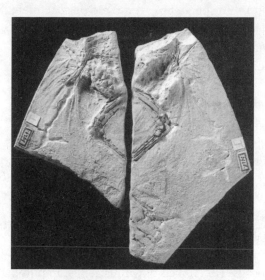

Figure 8. The Teyler specimen of *Archaeopteryx* was found in 1855, before any of the others, but was misidentified as a pterodactyl until John Ostrom reexamined the specimen in 1970.

"If it's not a pterosaur, *what is it?*" Ostrom thought to himself. He carried the slab over to the window where the light was better. In the next instant, the oblique sunlight illuminated the slab and brought up the impressions of feathers. Feathers would have been invisible in the harsh, even glare of fluorescent lights that equip any other modern museum in the world; feathers were the key that told Ostrom he was looking at an *Archaeopteryx*, not a pterodactyl (Figure 8).

"I knew what I was holding," says Ostrom, his face alight with the reminiscence. "Oh, I knew it"—here he gestures, striking his hands together—"I knew it like *that!*" Since 1861, there had been only the feather, the London specimen, the Berlin specimen, and the Maxberg specimen. And now there was one more, *his* specimen.

Ostrom remembers the ten minutes before van Regteren Altena returned as very long. His mind reeling with the magnitude of his discovery, he found himself in a terrible quandary during that hiatus. Should he tell his host what he had found? Would the curator allow him to take the specimen away for comparison and study if van Regteren Altena knew it was the fifth *Archaeopteryx*? Asking to borrow the fifth *Archaeopteryx* was a little more audacious than asking to borrow a specimen, even the type specimen, of an obscure species of pterodactyl that nobody cared about. Telling van Regteren Altena what the specimen was would clearly be the honest course of action, but a risky one. Ostrom had just made a great discovery and it might be taken from him if he told the truth. Maybe, Ostrom thought to himself, he should keep his mouth shut and simply ask to borrow the specimen. He could try to pretend that he hadn't realized what the specimen was until he had studied it further, *after* he had the specimen back at Yale. But the idea of deliberately deceiving his host and colleague was repugnant. It would be a most uncharitable act, one that was simply dishonest and selfish. No, he must tell van Regteren Altena. But then Ostrom wondered anew, aching with the conflict: was he willing to give up the glory of describing this new *Archaeopteryx* just to satisfy a point of honor?

"That ten-minute interval allowed my pulse rate to return to something approaching normal," Ostrom recalls. He choked down his selfish impulse. "I reached the decision that I had to tell him what he had, and then ask if I could borrow it." When van Regteren Altena returned, Ostrom showed him the feather impressions and told him that he believed this to be the fifth *Archaeopteryx*. Van Regteren Altena looked at it for a long minute, but said nothing at all. Then he picked up the slab and counterslab and disappeared into another part of the museum. Ostrom, stunned, could only watch him go and wonder what was happening. "I think at that instant he was in the same state of shock that I was ten minutes before. I remember sitting down on the chair that he had provided, thinking 'You blew it, John. You blew it!'" Ostrom sighs at the memory of his feelings at that moment. The triumph of morality seemed a little hollow, as he watched his great opportunity vanish with the specimen. But what was done was done.

"Anyway I went back to the rest of the collection that was available and was the reason I had come there in the first place. I could still make some observations on those specimens that had been identified as pterosaurs," Ostrom remembers, and pauses. "So about ten or fifteen minutes later, my host came back, and he was carrying a shoebox: a really battered old shoebox with a piece of string around it. And he presented it to me: 'Here, here, Professor Ostrom, you have made the Teyler Museum famous.' Those were his exact words. Didn't say yes, I could borrow it or anything like that, just 'Here.' The two pieces of limestone were in that shoebox."

In those few minutes, everything about *Archaeopteryx* had changed. Ostrom had found the fifth specimen and the fourth skeleton, not as impressive or complete as either the London or Berlin specimens, but no insignificant feat. And he had had a telling experience, one foretold by Huxley one hundred years before when he had remarked that a fossil bird, minus feather impressions, could not be distinguished from a reptile. Now Ostrom had lived through that flash of recognition, had felt for himself the thrill of looking at a certified reptile and seeing a bird.

There were other, important implications. Because it was found in 1855, six years before the feather or the London specimen, the Teyler specimen was truly the first *Archaeopteryx*. It had been living for 115 years under an alias, passing as a pterosaur. Awkwardly, this meant that the name given to the Teyler specimen, *Pterodactylus crassipes*, actually had priority, since it had been announced four years before von Meyer named the London specimen *Archaeopteryx lithographica*. Clearly the generic name *Pterodactylus* could not stand, since the specimen was patently not a pterodactyl, but the correct procedure under the rules of zoological nomenclature would be to attach the trivial name, *crassipes*, to the oldest valid generic name, *Archaeopteryx*. This would invalidate the well-known name of *Archaeopteryx lithographica*, which had been cited in the scientific and popular literature hundreds of times and which was, after all, an entirely appropriate name. After studying the specimen and finding it in no way different from the London specimen, Ostrom formally petitioned the International Commission on Zoological Nomenclature to make an exception in the

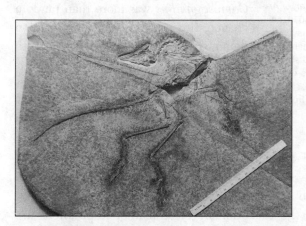

Figure 9. The Eichstätt specimen of *Archaeopteryx*, found in 1951, is wonderfully complete. However, it was misidentified as a small dinosaur, *Compsognathus*, until F. X. Mayr noticed the feather impressions some twenty years later.

normal rules, suppressing the name *crassipes* in favor of *Archaeopteryx lithographica*. In this, he was successful.

The discoveries were not over yet. Three years later, in 1973, F. X. Mayr, founder of the Jura Museum in Eichstätt, Germany, took a close look at a skeleton of the small dinosaur *Compsognathus* from Solnhofen. It was a less dramatic replay of Ostrom's experience, for Mayr realized that, once again, a skeleton of *Archaeopteryx* had been hiding under an assumed name. He sent the specimen off to Munich, where Peter Wellnhofer of the Bayerische Staatssammlung für Paläontologie und historische Geologie conducted a thorough study. The Eichstätt specimen, as it is known, does not have recognizable feather impressions although it is remarkably well-preserved otherwise (Figure 9).

Then, in 1988, Wellnhofer announced the finding of yet another *Archaeopteryx* skeleton, this time in the private collection of the former mayor of Solnhofen. The Solnhofen *Archaeopteryx* was again mislabeled—again as *Compsognathus*. The confusion of *Archaeopteryx* with *Compsognathus* was more than random chance; this persistent misidentification demonstrated a deep and resonant truth that Huxley had understood in 1868. It is that *Archaeopteryx* without feathers—or with feathers that had not yet been noticed—looks astonishingly like a small dinosaur similar to *Compsognathus*. *Archaeopteryx is* a feathered dinosaur. This sixth specimen is another beautiful fossil, largely complete save for the skull. Now that the specimen has been meticulously prepared (to the limited extent that preparation of any *Archaeopteryx* is possible), the exceptionally clear feather impressions are fully visible. It is the largest individual yet found, being about 10 percent bigger than the London specimen and almost twice the size of the Eichstätt specimen.

Figure 10. The Solnhofen specimen of *Archaeopteryx*, collected sometime in the 1960s, was sold to the former mayor of Solnhofen, Friedrich Müller. When the specimen was finally cleaned and studied, it was also at first misidentified as *Compsognathus*. Paleontologist Günter Viohl recognized it as *Archaeopteryx* from its limb proportions and from traces of feathers of the left wing.

In August of 1992, yet another *Archaeopteryx* was resurrected from the Solnhofen limestones. This seventh *Archaeopteryx* is known as the Solnhofer Aktien-Verein specimen (Figures 10 and 11); it was recognized at the Langenaltheimer quarry as soon as the slab was split. It provided three surprises, new features that significantly altered the view of *Archaeopteryx*. The skeleton is very small—even smaller than the Eichstätt specimen—but apparently adult. This has led Peter Wellnhofer, the German paleontologist who described it, to suggest that it represents a new species, *Archaeopteryx bavarica.* Apart from size, the specimen includes two new anatomical features: a bony sternum, which is unknown in any other specimen and is crucial for the attachment of flying muscles; and a set of bony, interdental plates that are preserved on the inner side of each lower jaw between all tooth positions. Similar interdental plates are found in two different groups that have been suggested as possible ancestors for *Archaeopteryx:* the theropod dinosaurs and a still more primitive reptilian group, the thecodonts.

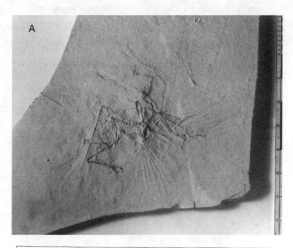

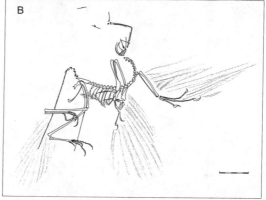

Figure 11. Found in 1992, the most recently discovered *Archaeopteryx*, shown in a photograph (A) and a drawing (B), is known as the Solnhofer Aktien-Verein specimen. Paleontologist Peter Wellnhofer, who described the fossil, suggests it is a new species, *Archaeopteryx bavarica.*

Seven skeletons, one feather: that's all there is to tell us of this elusive creature. Nearly half of all known skeletons were initially misidentified, one mistaken for a pterosaur, two others mistaken for the small dinosaur *Compsognathus.* These few fragile specimens have borne a heavy burden.

They include the most beautiful fossil in the world; the most famous fossil in the world; the jewel in the crown of the most beautiful museum in Europe; the proof of evolutionary theory; the evidence of the origin of birds; the secret of bird flight. How can seven scant specimens be so significant? How can it be otherwise?

Archaeopteryx derives immense power from all of this—strength to capture the imagination, fuel to ignite the intellect, heat to sear the emotions. To be *Archaeopteryx* is to be extraordinary; there is no other choice. To touch *Archaeopteryx*—to study it, handle it, think of it—is to succumb to its power and to enter the debate.

Chapter 2.

What's the Flap?

I went to see John Ostrom to find out what the arguments were all about. A smallish, silver-haired man, Ostrom is now retired from the Yale Peabody Museum. Once quick of movement, he has slowed down a little in his walk and in his speech; he wears a hearing aid in each ear, and he sometimes asks visitors to repeat a question, not certain he has heard it exactly right. But when he talks about the origins of bird flight, the hesitancy disappears and the intelligence and articulateness that have made him so successful become obvious. His face lights up with curiosity and passion; a younger Ostrom appears. Whether he is right or wrong about the origins of bird flight, his enormous impact on his field is unchallenged.

"He is The Man," explains David Weishampel, a paleontologist at the Johns Hopkins Medical School. "He was the one who inspired the next two generations of paleontologists to do their science in a whole new way." That new approach was to study ancient life forms as a part of biology. Due to the unique perspective pushed by Ostrom and others of his generation, modern paleontologists no longer think of fossils as lumps of petrified bone, dry yet fascinating to an esoteric few. Today's paleontologists seek to glimpse the real creatures behind the bones: the ones who moved, vocalized, squabbled over nesting or breeding sites, used colorful displays or courtship dances to attract mates, and preferred certain habitats to others. It is a notion based on a strong sense of the tremendous continuity of life, throughout the millennia, and fueled by the creativity to invent new ways to extract biological information from the traces that remain.

The rise of paleobiology—the movement in which Ostrom's influence has been felt—is one of three great themes of modern paleontology. The other two innovations mark the last few decades of work. They are the revolution in the methods of classification (the rise of Hennigian cladistics) and the development of a highly quantitative and technical approach to understanding the links among anatomy, physiology, and function. Cladistics and quantitative biomechanics are not Ostrom's forte. He acknowledges their value but has never embraced these approaches fully. Yet Ostrom's almost naïve wonder at the magnificence of life and evolution does more than any methodological breakthrough can: it brings the past to life. Ostrom inspires.

"This issue of bird flight has to be, in my mind, one of the most intriguing questions in vertebrate evolution that one could venture into," he explains. "Why? What is it that cultivates perfectly normal, ground-living creatures into flying up into the air? *Why?*" His enthusiasm is compelling. In his presence, you realize that paleontology provides a great deal of *joy* to its practitioners. It is this quality in Ostrom that has made him a much-beloved figure in paleontology. In 1994, he was awarded the Romer-Simpson Medal, the highest honor bestowed by the Society of Vertebrate Paleontology. Even those who disagree with his work and theories are reluctant to criticize him too harshly.

The questions he started with—the issues that Ostrom almost single-handedly turned into one of the great scientific controversies of the twentieth century—are why and how bird flight evolved. Logically, the first step toward answering these questions is figuring out how *anything* flies. What are the prerequisites of flight and what aspects of a creature's anatomy and structure enable it to fly? What features make one species an excellent flier and another a marginal one? Aerodynamic theory and the successful invention of aircraft notwithstanding, these have proven remarkably difficult questions to answer clearly.

Humans have been trying to discover the secrets of flight since the thirteenth century, if not earlier. In about 1500 A.D., Leonardo da Vinci was persuaded that studying birds would yield the key to flight. He kept meticulous notes on his dissections of birds and his observations of their flight, writing, "To attain to the true science of the movement of birds in the air, it is necessary to give first the science of the winds. . . . You will study the anatomy of the wing of a bird, together with the muscles of the breast which are the movers of these wings." The result was Leonardo's sketches of flying machines and winglike devices that are strangely prescient. Yet despite his earnest efforts and unquestioned genius, Leonardo was unable to crack the elusive problem of the principles of aerodynamics. About one hundred years later, John Wilkins (one of the founders of the Royal Society and Lord Bishop of Chester) outlined the four ways that a man might fly: 1) with the help of angels; 2) with the help of fowls; 3) with wings fastened to his body; and 4) in a "flying chariot," meaning, presumably, a device or vehicle powered by something other than human muscular effort. Wilkins judged the first three rather improbable, and in this he was largely correct. Birds and angels are strangely unwilling to assist humans in flying, while attempts to fasten wings to humans are notoriously dangerous.

It was not until December 17, 1903—four hundred years after Leonardo—that Wilbur and Orville Wright made their epochal flight on the beach at Kitty Hawk, North Carolina. Their flying machine succeeded in staying aloft for a full twelve seconds, covering 120 feet. Quantitatively, it is a pathetic performance for the achievement remembered as the ultimate demonstration of human mastery of the air, yet even this modest triumph was the culmination of hundreds of years of human endeavor and ingenuity. Indeed, the history of the development of human flight provides an instructive framework for thinking about the evolution of avian flight, because humans were forced to analyze and evaluate the separate components of flight in order to master them.

The basic problem—the solving of which makes birds and airplanes so magical—is that of conquering a hostile medium, the air. Gravity is an implacable force and the air a poorly supportive environment. Still, flying is one of the most universal themes of human dreams, myths, and stories. Its symbolic power lies in the defiance of gravity. Fliers literally rise above the mundane, earthbound existence to which the rest of us are confined. Thus, long before humans achieved any sort of flight, many cultures developed religious beliefs featuring gods or celestial beings who could fly. Some of these mythical beings had wings, like angels or the deities of Egypt, Minoa, and Syria; others simply flew magically, without any special anatomical apparatus. One of the best-known is the Greek tale of Icarus and his father, Daedalus. It is a complex story of fathers and sons, of pride and excessive ambition, of talent and tragedy. These elements are woven together into a structure that has been repeated and adapted numerous times in Western culture. The characters of Daedalus and Icarus, and all that they symbolize, appear in everything from the brilliant novels of James Joyce to the haunting sculptures of Michael Ayrton. Before investigating the origins of flight, I had forgotten all but the most basic plot line of the Greek myth, but there is much that makes this classic story relevant to the scientific search for the origins of flight.

Daedalus was a sculptor and an architect. His name reflects his extraordinary talent; various translations of Daedalus are "the skillful one," "the cunning fabricator," or "the craftsman of the gods." Because he was so clever at making things, Daedalus was hired by King Minos of Crete to build an intricate and inescapable labyrinth, which the king needed as a consequence of his own sins. Minos had been sent a snow-white bull for sacrifice by the god Poseidon, who had helped Minos to obtain his throne; rather than fulfill his obligations, Minos kept the magnificent animal alive instead. Poseidon punished Minos by causing Minos's wife, Pasiphae, to fall in love with the bull. From their union, she bore a monstrous male child, a half-bull, half-human known as the Minotaur. The labyrinth was to be the Minotaur's prison, a means of concealing and containing the tangible evidence of shame and wrongdoing that this unnatural son represented. But when Daedalus overstepped his bounds with the king, the king in turn devised a punishment for Daedalus as cunning and painful as the one Poseidon had meted out to him. Both Daedalus and his son, Icarus, were shut into the labyrinth with the deadly Minotaur. Thus the clever fabricator and his most beloved "product," his son, were imprisoned within the walls of Daedalus's own creation, which he had made inescapable. Despite the labyrinth's design, Daedalus managed to contrive their escape by fashioning wings of feathers and wax for himself and his son. Tragically, Icarus flew too near the sun. The wax melted, the wings disintegrated, and the miraculous escape ended in Icarus's death. Daedalus's pride and excessive cleverness first saved and then doomed his own son. Like many myths, this one is designed to be in part instructive, to teach the listener how to behave appropriately in the society from which the story springs. The sobering moral concerns knowing one's place and limiting one's aspirations: fly too high— too near the sun—and you are surely doomed. Icarus's flight is a metaphorical invasion of the territory of the gods, a symbolic repetition of his father's audacity in making a prison for a half-god, the Minotaur, and wings for mere mortals.

Aside from the immense power of the story, what struck me is the choice of flying as an integral element of the plot. Here is an explicit demonstration of the depth and antiquity of our feeling that the air is a dangerous place for the likes of us. Even today, our ancient suspicions of technology adulterate our pride at the human expertise that lets us invent ways to invade formerly unknown domains and realms. More than the bicycle or the automobile—both vaunted inventions of their day—the technical accomplishment of human flight is an especially valued achievement because it resonates with centuries of accumulated myths and stories like the one of Icarus and Daedalus. The potency of the symbolic power of flight is reflected in our cultural reactions. Starting in the 1700s, an avid cult grew up around the early aviators, and pilots continued to enjoy a celebrity status until air travel became conventional and routine in the mid-twentieth century. The same hero worship that was lavished on the Wright brothers, Charles Lindbergh, Amelia Earhart, and others was then transferred to the wartime fighter pilots and flying aces and later on to the daredevils like Chuck Yeager, the first man to break the sound barrier. Later still, astronauts replaced pilots as the fabled explorers of the hostile and unforgiving atmosphere above us. In the phrase of Thomas Wolfe, astronauts had the Right (or Wright) Stuff. They were heroes.

Paleontologists like John Ostrom make the link from today's technological heroes back into the depths of time, millions and millions of years ago, to a much earlier evolutionary triumph. Insects, reptiles, bats, and birds conquered the air long before humans ever appeared—and did it over slow generations, by the dumb mechanism of undirected natural selection. Ostrom's wonder and curiosity brought me back once again to the beginning, to the stunning evolutionary accomplishments of the original fliers, the strange and wonderful likes of pterosaurs and *Archaeopteryx*. These first invasions of the air were even more amazing than the human ones. On a pragmatic level, flying is a very complicated task. The first challenge to be solved is not the social or moral one of honoring the gods, knowing one's place, and avoiding punishment. The task at hand is a practical matter of finding a way to leave the earth for the air: takeoff. And for birds, insects, bats, and pterosaurs, the equipment had to be evolved from existing structures, not manufactured.

Takeoff requires harnessing some force to create enough lift to support the body. In its simplest form, lift is created by forward movement combined with an airfoil or wing of suitable shape. In cross-section, the archetypal wing looks like this: convex on the top and concave below (Figure 12). In moving air—produced either by propelling the airfoil forward or by placing it in a

naturally occurring air current—
this shape makes the air passing
over the top of the wing move faster
than that passing beneath the wing.
The difference in air speed in turn
creates a pressure differential between the top and bottom surfaces
of the wing that, when it is great
enough, actually lifts the body (or
plane) upward. Thus two elements
are required for takeoff: a wing or
airfoil of appropriate shape, and
some form of energy that can be
used to produce speed and lift.

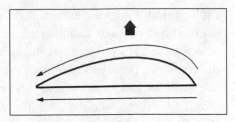

Figure 12. An airfoil produces lift because in
cross-section its curved upper surface is longer
than its flat lower surface. This asymmetry
means that the air moving over the upper surface must travel farther (and hence faster) than
the air that moves under the lower surface, creating a pressure differential that results in upward lift (indicated by the broad arrow).

What sources of power or energy can be used? The simplest case is perhaps that of a kite. With a kite, the design of the airfoil is so favorable—the
surface area of the wing is so enormous compared to its weight—that a mild
breeze or a brief run on the part of the kite flier is usually sufficient to propel the kite aloft. Kites require only a marginal input of muscular power for
takeoff, above and beyond the wind power that keeps the kite aloft once it is
launched. Kites are an old invention, widely known prior to the seventeenth
century; huge observation kites were used in the American Civil War in the
1860s to spy out the enemy's position. Although these kites were large
enough to take a human aloft, even the use of manned kites revealed little
about how to move a substantial weight, like a human, freely and at will
around the sky. Takeoff was solved after a fashion, but not flying.

Rather than using wind power and an enormous airfoil, the first viable
attempts at human flight used a balloon and hot air. The year was 1783, the
inventors were Joseph and Étienne Montgolfier, and the place, southwest-
ern France. The theory behind their invention was interesting. The pair
had astutely observed that smoke rises in the chimney; from this, they deduced that it possessed an enigmatic property that they called "levity." (Of
course, what they had discovered was the principle that hot air rises; smoke
in and of itself does not possess a quality known as levity, although it is a
charming notion.) The solution to the problem of takeoff, as they saw it, was
one of capturing sufficient levity in a lightweight container. They invented
the balloon, using at first a small silk bag that they held over the fire until it
filled and was carried upward. Heartened by this success, they constructed

a "Globe Aerostatique," a spherical bag, about thirty feet in diameter, of lightweight cloth backed with paper. They invited the public and local dignitaries to witness their first attempt to fly the globe on June 5, 1783. The brothers held it over a smoky fire and, to the amazement of the observers, the globe inflated quickly, floating up a considerable distance in the air (reported to be over a mile) and traveling a distance of a mile and a half before it touched down. There were no passengers on this initial flight; it was too dangerous. But the potential for carrying humans, animals, or freight on such a globe or aerostat was apparent. The Montgolfiers were summoned to speak to the Academy of Sciences in Paris on their remarkable flying machine, at a meeting attended by Benjamin Franklin among others.

In a manner typical of the history of aviation, the Montgolfiers were only a short distance ahead of their rivals. Almost as soon as the Montgolfiers' aerostat was announced, a team that comprised physicist J.-A.-C. Charles and another set of brothers, Anne-Jean and Nicolas-Louis Robert, made a balloon of varnished silk. Ignorant of the Montgolfiers' simple plan for filling their aerostat with levity, Charles's team decided to use a newly discovered gas, hydrogen. Being lighter than air, hydrogen's natural propensity was to rise, thus producing the lift needed for takeoff. Their balloon, like that of the Montgolfiers, was highly successful and gave rise to the nickname "Charlière" for any hydrogen balloon in France for years to come. Their very first Charlière took off on the afternoon of August 27, 1783, in Paris. Witnessed by a delighted crowd of enthusiasts, the Charlière rose and flew some fifteen miles before coming down in the rural village of Gonesse, where the success of the flight was somewhat mitigated by the reactions of the peasants and laborers. Frightened by this horrible apparition that dropped out of the sky, they rushed out and "killed" the monster with their scythes and pitchforks.

In the next month, Joseph Montgolfier addressed the Academy of Sciences and subsequently was asked to demonstrate his aerostat for King Louis XVI at his palace of Versailles. On this royal occasion, a sheep, and, ironically, two birds (a rooster and a duck) were unwilling passengers on the flight. Their survival seemed to show that high-altitude travel posed no undue health hazards, so plans were quickly made for a manned flight. A new and larger aerostat was built, with a circular gallery suspended beneath the balloon where passengers had to maintain a smoky fire (to generate levity) throughout the flight. The initial suggestion—to use a pair of condemned prisoners as the first passengers, in case anything went wrong—was scorned as unworthy. The moment was too important and the accomplishment too triumphant to be marred by the presence of such men. Instead, Jean-François Pilâtre and François Laurent, marquis d'Arlandes, bravely volunteered for the ascent. On November 21, 1783—only a few months after the invention of the hot-air balloon—Pilâtre and Laurent embarked on their adventure, remaining aloft for twenty-three minutes and covering a distance of perhaps ten miles before they landed safely. It was the first documented manned flight in the history of the world. Further experiments with manned balloon flights soon demonstrated the superiority of the hydrogen to the hot-air balloon, in that the former was able to stay up without the considerable effort of continuously stoking the fire. The dangerous vulnerability of hydrogen balloons to explosion, due to the extreme volatility of the gas, was not appreciated until much later. But after the German airship *Hindenburg* exploded into a much-publicized fiery ball in 1937, hydrogen balloons were largely abandoned.

Both hydrogen and hot air were capable of providing the necessary lift for takeoff, but their difficulties pointed up the next two problems in flight: manouvering or propelling the craft once it was airborne and controlling the landing. Balloons are sadly subject to the vagaries of prevailing winds, as many balloonists have found to their regret. Basically, a balloon gets you *up*, but leaves you largely unable to steer or control your flight path effectively. Eighteenth-century balloonists tried many possible solutions—sails, oars, paddle wheels—with no success. The spherical shape of a balloon designed to capture levity or lighter-than-air hydrogen was all wrong for directed movement once the passive takeoff had been accomplished. An elongate shape, reminiscent of a fat cigar, would steer better but the problem of devising a lightweight power plant to propel the balloon became apparent. In a sense, progress in ballooning was stymied once again by the need to find a source of sufficient power, only now the power was needed for propulsion and maneuvering, not for ascent.

The prerequisites of powered, man-made flying machines are that the source of energy be both compact and lightweight. Human muscle power is simply inadequate given our anatomy, despite Icarus's mythical flight and fanciful films like *Brewster McCloud,* in which a young man straps a wing-like contraption to his body and learns to fly by doing something rather like the butterfly stroke in air. If humans are too feeble—or too inefficiently built—to succeed in flapping flight, how then do small and fragile-seeming birds manage with muscle power alone?

54

The answer is "with difficulty." The evolution of flight has honed avian anatomy into an extreme and remarkably adaptive configuration. The most obvious adaptation involves weight-saving: the lighter the body, the easier it is to fly. Modern bird bones are hollow, filled with special air sacs that are connected to the lungs and form an ingenious system that helps move air through the bird's complicated respiratory apparatus. Hollow bones reflect the need to compromise strength and lightness, for it is essential that bird bones be able to withstand a certain amount of buffeting and the inevitable collisions with branches, house windows, or other obstacles. Part of the protection is afforded by the geometry of the bones. One of the axioms of beam theory is that a fat cylinder (or near-cylinder, like the long bones of the leg or wing) with narrow walls can be mechanically as strong as a narrow cylinder with thick walls if the substance that makes up the walls is properly arranged. In bones, the walls comprised bone tissue itself, while the central cavity is filled either with marrow or, in birds, with air sacs that are an integral part of the ventilation system. Long bones typically have a slightly oval rather than a round cross-section; the trick is to align this extra tissue with the axis along which the greatest stresses will be applied. Under these conditions, a bone shaped like a hollow near-cylinder can be extremely strong and resistant to breaking. Pterosaurs and birds probably evolved hollow bones independently but for similar aerodynamic reasons; the bones of *Archaeopteryx* are hollow, but the extent of air sacs is difficult to assess.

Additional protection is afforded by the feathers themselves. The original function of feathers—why they evolved—is a controversial subject that I take up later in this book. Whatever their original function, feathers now offer four vital advantages to birds. First, feathers help birds maintain a roughly constant internal temperature, no matter what the ambient temperature may be and no matter what activities they are engaged in. Second, feathers form an incredible airfoil, thickening the profile of the cross-section of the wing to a highly efficient shape while adding almost no weight. Feathered wings are much more efficient airfoils than membranous wings, like those of a bat or pterodactyl. Third, feathers are also sufficiently mobile to act as slots or flaps: antistalling mechanisms that are terribly useful for landing. Slots are created in birds' wings by the separation of the tips of the primary feathers, like the fingers of an outstretched hand. Slots are especially crucial in birds like albatrosses or condors that seem to be approaching the maximum weight for a flying animal. Finally, feathers are light, stiff, and strong, forming an excellent shield that protects the lightweight bones and the thin skin of the wing from damage. Because a coat of feathers comprises many individual, overlapping units, damage to any particular section of the feathers is unimportant. Losing a few feathers to a cat or a thorny branch is far less dangerous to a bird than tearing a membrane would be to a bat. Lost feathers can be replaced whereas damaged membranes, even if repaired, may be scarred and their function compromised. What's more, the multiple, overlapping layers of feathers are tough and resilient. Ancient Japanese feather armor was more than a ceremonial whimsy; feather armor protected its wearers remarkably well against swords and daggers. In contrast, the traditional metal suits of armor developed in Europe that we are used to thinking of as so effective were in fact dangerously cumbersome, unwieldy, and enormously heavy. While a mounted knight in armor was difficult to injure, mounting itself was nearly impossible. Knights in armor had to use special derricks to lift them up onto their horses. One of the greatest dangers to an armored knight lay in being unmounted by an opponent, who then had the downed knight at his mercy. Chain mail was viewed as an improvement because it was lighter in weight than armor, even if the protection it offered was less complete.

Although birds have all of these special anatomical adaptations for flight, the question still remains: how exactly do avian muscles propel the bird through the air? Several evolutionary changes in the bones and joints of the wing—deviations from the more generalized arm structures that must have been the precursor to wings—are key. Birds generate lift by flapping their wings, thrusting down and slightly forward against the air resistance. This is the downstroke or power stroke, the obvious part of flapping flight. The other movement—recovery or upstroke—is more subtle. If the wing's shape were held constant and flapped downward and then upward, both the downstroke and the upstroke would push against the air. With a constant wing shape, the upstroke would propel the bird downward just as strongly as the downstroke moved it upward. The net result would be no lift and no forward movement. To avoid this problematic fate, birds use one of two strategies. Some birds slightly fold and flex the wings on the upstroke, effectively reducing the area of the airfoil and thus minimizing the downward thrust that is generated. Thus these birds push down with a large wing and push up with a small one. Other birds keep their wings outspread but tilt their wings differently in different segments of the wingbeat, thus adjusting the effective surface area.

Another mechanism that passively helps to create an asymmetry between the upstroke and downstroke involves the structure of flight feathers. In these feathers, the rachis—the stiff quill—divides each feather into two asymmetrical vanes, with the forward or leading edge vane being narrower. During the downstroke, the pressure on the underside of the feather causes it to twist along the longitudinal axis of the rachis. The result is that the trailing edge vane of each feather is pressed upward, making the entire wing a broad structure that pushes against the air without allowing air to pass between adjacent feathers. In the upstroke, the opposite occurs. The pressure on the upper surface of the wing twists each feather around the rachis so that the leading edge vane is moved upward. This movement separates adjacent feathers from each other slightly, forming a tiny slot through which air can pass. Thus both the movement of the wing itself, in opening and closing, and the movements of individual feathers act to improve the efficiency of the wingbeat.

A familiar analogy that may help explain this movement clearly depends upon the functional similarity between flapping and rowing a boat. In rowing, exactly the same problem pertains: since the oar must move both backward and forward, the oarsman has to find a way to propel the boat forward during the power part of the stroke without sending the boat an equal distance backward with the recovery segment of each stroke. The power part of the oarsman's stroke is the equivalent of the bird wing's downstroke; it provides thrust resulting in forward (or in a bird, forward and upward) movement. The oarsman's recovery takes the place of the bird's upstroke and ideally produces no backward movement at all. Unlike birds, however, a human oarsman generally sits facing away from the direction of travel. This is logical because humans can exert more power as they pull the handles of the oars toward their chest by contracting their arms. Pushing the handles away, by extending the arms, is a weaker movement. But when the handles come toward the rower's chest, the paddles on the other end of the oars move in the opposite direction, pushing against the water. The result is that the boat is propelled in the same direction that the oar handles are moving, using the more powerful movement. Even as some birds twist their wings during the recovery stroke, so the oarsman uses a series of wrist motions to change the orientation of the oar's blade relative to the water. Like a wing, the blade has a broad, flat surface and a narrow cross-section.

The rowing stroke starts with a moment known as the catch, the entry of the blade into the water. The blade must be perpendicular to the water's surface (or vertical) at the catch, so that the thin cross-section passes easily into the water and the blade is positioned to generate thrust. As the oarsman contracts his arm muscles, he also uses his torso and back muscles to stabilize his position. In a rowing scull, each rower sits on a sliding seat and is responsible for moving a single oar. The rower contracts his arm muscles while he also extends his leg muscles, straightening his knees. (Human legs work most powerfully in extending the knee, while human arms work most powerfully in bending the elbow.) During this coordinated movement, the oar is pulled through the water with the blade set in a vertical position, so that its maximum surface area is perpendicular to the direction of movement of the oar. Pushing against the water with the broad surface of the blade thrusts the boat forward. Once the oar has reached its maximum excursion, the recovery stroke begins. The first movement is to "feather" the oar. The rower maintains a constant grip on the oar handle, but moves his wrist to rotate the oar until the blade is in a horizontal position. Once the blade is horizontal, the rower lifts the oar out of the water. (Though the movement of the oar is similar to that performed by birds' wings, birds generally do not take advantage of the difference in density between air and water to minimize the thrust generated in recovery. To do this, birds would have to fly underwater, as penguins do, and make their recovery with the wings in the air, which penguins do not.)

In rowing, the aim of the recovery is to return the blade to its starting position while minimizing both the energy needed to accomplish this task and the backward thrust potentially produced by moving the oar. The most important aspect of recovery is lifting the oar blade out of the water, since air offers much less resistance than water. This is why flying fish "fly" out of the water. They perform no propulsive strokes while in the air but rather glide using thrust that was produced by movements performed while they were under the water. Why do flying fish fly? Because it works. For the same thrust, a flying fish can travel farther in air than in water because air has a lower density than water; in air, less of the energy put into the stroke is wasted in friction or drag when the fish is in the air, so the glide is longer. Rowers also take advantage of their ability to feather the oar, which means that the oar travels through the air oriented so that only its thin cross-section, rather than its large surface area, faces forward. This lessens the friction or drag produced by the oar as it moves through the air and thus minimizes energy expenditure further. At the end of the recovery the rower "squares" the oar, rotating the blade back into the vertical position in preparation for the catch.

In rowing, as in flying, the power source is muscular. But humans gain an enormous advantage in rowing by lifting the oar out of the water and into the less dense air during the recovery, a strategy which is moot in flying (except for the case of flying fish). Instead, the anatomy of birds has been substantially altered by natural selection to create a structure in which muscle power generated by the forelimb works more effectively than it does in humans. Bird wings differ from human arms (or any more typical mammal's forelimbs) in a number of crucial ways. Starting from the hand and working up to the shoulder, even a cursory inspection of avian anatomy reveals that the number of fingers has been reduced. The generalized mammalian hand has five fingers, which anatomists and embryologists refer to by roman numerals from I (the thumb) to V (the pinky). This has been evolutionarily modified in a bird so that there is one main finger at the central axis of the hand, with a much reduced digit on either side. Traditionally, paleontologists and embryologists have disagreed on the question of which specific three digits have been retained in birds; embryologists favor II–IV and paleontologists favor I–III. In either case, only three fingers are represented on the avian hand and the primary flight feathers, or remiges, are attached directly to them. Moving from the fingers to the wrist, additional evolutionary changes are apparent. The nine wrist bones sported by generalized mammals have been reduced and fused in birds. Three of the mammalian wrist bones are fused into one, the avian carpometacarpus, while the other six are fused into two avian wrist bones called the radiale and ulnare. In the forearm, both birds and mammals have two bones, the radius and ulna, while there is one in the upper arm, the humerus. This structural pattern is so important to flying that, despite the many species of birds in the world, the wing structure is nearly invariant except in those species that have abandoned flight altogether.

Not only is the structure of the forelimb or wing distinctive, its action is specialized for flight, too. In a human, nearly every joint in your forelimb can be flexed separately. The shoulder can be flexed without flexing the elbow, the elbow can be flexed without flexing either the wrist or the shoulder, the wrist can be flexed without flexing the fingers, elbow, or shoulder, and the joints within the fingers can be flexed without flexing any other joints. Only the joints within each digit, marked by knuckles, are arranged so that the same muscle and tendon can operate all of the joints within one finger simultaneously. Some people have sufficient control that they are able to flex only their most distal knuckle joint, bending the last joint and moving the distal phalanx without bending the entire finger, but this ability is relatively unusual. The point of arranging the forelimb in humans and many other mammals in this way is to maximize the flexibility of the forelimb and the complexity of its possible movements.

In contrast, the elbow and wrist joints in bird wings are mechanically linked so that their movements are inevitably coordinated. Extending the elbow automatically extends the wrist and thus the rest of the wing; flexing the elbow automatically folds the wrist, flexing it sideways (away from the thumb) in a motion impossible for humans, who can flex their wrists only up and down. This coordination is the result of both muscle action and bony arrangement. At any joint, there are two primary movements: flexion, causing the bones of the joint to assume a more acute angle; and extension, straightening the joint so that the bones form a 180-degree angle. When muscles like the avian biceps contract, the elbow is flexed and the two bones in the forearm are brought closer to the humerus in the upper arm. To this extent, the action is very similar to the result of contracting the biceps in a human. But in birds, the articulation between the radius (the more slender bone of the forearm) and the ulna (the stouter bone of the forearm) and the humerus is special. The radius and ulna each articulates with its own condyle or bony bump on the distal end of the humerus at the elbow. However, the condyle for the radius is bigger—a more pronounced lump—than that for the ulna, especially in the position during flexion. Thus flexing the elbow moves the head of the radius up onto the most prominent part of the condyle. The radius and ulna always lie side by side, but when the radius moves up onto the most prominent part of the condyle, it is shifted longitudinally toward the wrist. The distal end of the radius in turn pushes against the various carpal or wrist bones, which themselves push against the metacarpals (the equivalent of the bones that underlie our palms). As the bones of the wrist and hand move in response to the shove from the radius, they automatically cause the wrist to flex, too. During extension of the elbow, the opposite action of the radius occurs; it shifts toward the elbow, causing the carpal bones and the metacarpals to return to their starting position, thus extending the wrist. A single muscle (the extensor metacarpus radialis) that in mammals extends only the wrist can thus also unfold the elbow and the entire wing in birds. In other words, the radius acts as a mechanical connecting rod that coordinates actions between the elbow and wrist. In more humble terms, the wing can be said to operate somewhat like an old-fashioned device once found in every corner grocery store that was used to take cans down from high shelves (Figure 13). Flexing both elbow and wrist automatically tucks the wing up against the body, out of harm's way; extending both joints places the wing in a position to push downward and forward in order to generate thrust and lift.

Some of the greatest evolutionary differences become obvious when I compare the shoulders of birds and mammals, as other anatomists have done before me. The mammalian shoulder blade, or scapula, is a roughly

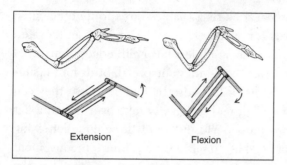

Extension   Flexion

Figure 13. The wing of a bird has a mechanical linkage system that coordinates the automatic extension or flexion of all joints in the wing. Thus birds can spread or furl their wings with a minimum of muscular effort.

triangular-shaped bone with a socket to receive the head of the humerus. The mammalian scapula also articulates with the collar bone or clavicle on its side of the body, via a bony process known as the coracoid that sticks out from the scapula. At the midline of the body, the right and left clavicles meet the sternum or breastbone, a fairly flat bone that is roughly rectangular in outline, with the long sides running vertically. From the top to the bottom of the sternum's long sides, there are articulations for the clavicles and, below them, articulations for the pairs of ribs. In contrast, the scapula in birds is elongate and straplike, not triangular, although it still lies along the ribs on the back of the bird. The coracoid is an entirely separate bone, not a process sticking out from the scapula. Instead of articulating only with the clavicle, the avian coracoid joins the scapula in forming the socket for the head of the humerus. The right and left clavicles of mammals are fused in birds into a single V-shaped bone known as the furcula, or wishbone. The avian sternum is greatly enlarged compared to the long, flat bones found in mammals. The sternum is a large, keeled, bony structure known as the carina. Birds need a carina because their chest muscles are also flight muscles that pull the wings down and forward. Flight requires such powerful strokes that the avian flight muscles are very large and need additional bony areas to attach to. The attachment areas at the other end of the flight muscles, on the humerus, are also enlarged with extra bone. Almost all of the chest muscles involved in flight arise from the carina; they are larger in powerful fliers like pheasants and smaller in poor fliers, like steamer ducks, or flightless birds like the kiwi. The size and strength of the carina also vary with flight ability, being much smaller in flightless or poorly flighted species and much bigger in powerful fliers. In fact, in flightless birds the carina may be cartilaginous, rather than bony, or even completely absent. Clearly, the elaboration and specialization of the shoulder girdle in birds is a response to the demands of flight, specifically to the need for large and strong bony anchors to which flight muscles can attach.

There is one more anatomical trick that enables birds to fly efficiently: the wing flip. Takeoff for birds (or anyone else) is easier from an elevated substrate than from the ground. As hang-gliders favor cliff edges or artificial ramps that they can run off of with their wings outspread, birds find it simpler to take off from trees or cliffs or high-rise buildings, where they can simply spread their wings, push off, and let gravity take over. From a flat surface like the ground, a bird must perform two or three complete wing strokes in order to generate enough lift to move itself upward. Inconvenient it may be, but practically speaking takeoff must occasionally occur from the ground. To do this, the bird must complete a successful upstroke that lifts its wings upward and backward until they are higher than the level of its back, followed by a downstroke that forcibly flaps them downward. The downward action, produced by the large breast muscles anchored to the carina, is no problem. Contracting those muscles moves the bones on the mobile end (the wing) closer to the relatively fixed end (the carina). But how does a bird manage to elevate its wings above the level of the point of attachment of any of its muscles? What possible leverage can any muscle exert to accomplish this feat? None, unless the bones and muscles of the wing are arranged so that a wing flip is possible (Figure 14).

The muscle that is primarily responsible for the wing flip is called the supracoracoideus. It arises from a large area on the keeled carina and the sternum itself. The muscle fibers pass upward, toward the shoulder, and converge into a single, strong tendon that passes through a bony tunnel (formed by parts of the coracoid, scapula, and furcula) known as the triossial canal. From the triossial canal, which is the highest point of the shoulder, the tendon turns to run downward again to insert on the back surface of the upper end of the humerus. Fundamentally, the triossial canal operates as a pulley mechanism. When the supracoracoideus muscle is contracted, the insertion of the muscle on the proximal end of the humerus is pulled closer to the opening of the triossial canal. As a result, the entire humerus is abruptly flipped upward and outward, as if it

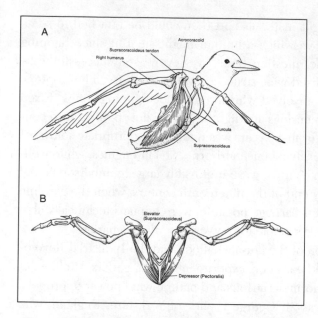

Figure 14. The muscle that causes the wing flip in birds is the supracoracoideus, which originates on the keel of the sternum and inserts via a tendon on the humerus or upper bone of the wing. Flexion of the muscle brings the humerus, and the rest of the wing, into a vertical position above the bird's back to start the downstroke. (A) shows the position of the muscle in a whole bird; (B) shows the muscle's action in a head-on view.

were standing on its head. Inevitably, moving the head or proximal end toward the canal flips the distal or elbow end of the humerus up into the air, which places it above the back. As the wing reaches maximum height, the extensors of the wing come into play, so that a fully extended wing is now in position above the bird's back, ready for the downward flap.

Humans can't do this for a variety of reasons. First, takeoff is even more difficult for us than for birds because our bodies are dense and heavy; our bones have thick walls and are filled with marrow, not air sacs. Second, our chest muscles operate at a mechanical disadvantage relative to those of a bird. We lack the triossial canal and the special anatomical arrangement of the shoulder girdle that makes the wing flip possible. Third, our chest muscles and sternum are far too small and feeble to power our arms in effective flaps. Our chest muscles would have to be proportionately much more powerful than a bird's to produce a powerful enough stroke to fly, especially given our heavy bodies and the mechanical disadvantages of the structural arrangement of our bones and muscles. Fourth, and finally, our arms are not broadened into effective airfoils. In other words, we lack all of the major elements—wings, efficient shoulders, and requisite muscle power—to flap effectively.

The lack of appropriate power was probably the greatest stumbling block to human flight. Since balloons were too difficult to control once aloft, an alternative approach was tried: making devices with flapping wings based on birds, bats, or pterodactyls (ornithopters). These were relatively simple to build out of lightweight materials and then could be attached to some sort of harness or chair in which a human might sit. But since flapping proved to be out of the question (and has never been successfully attempted by humans), a fixed-wing structure was needed. Even helicopters, a much later invention, involved whirling and not flapping wings. Fixed wings of appropriate dimensions and lightness were not so difficult to construct, but the problem for many years was finding an appropriate source of power for takeoff and for flying once airborne. Steam engines, while well-known in the nineteenth century, were impossibly large, cumbersome, and heavy. Not until the very end of the nineteenth century, when the gasoline engine made the horseless carriage possible, were humans technically able to make engines that were suitable for powering a "flying chariot."

Between the invention of the hot-air balloon and the Industrial Revolution, a number of would-be aviators experimented with gliders, kitelike devices that relied upon enormous airfoils and natural wind power to propel a human. Among them was Sir George Cayley of Yorkshire, England, who pursued the question of flight for more than sixty years (1792–1857) through experiments with models and full-size gliders. He cast the basic problem as being "to make a surface support a given weight by the application of sufficient power to overcome the resistance of the air" and eventually worked out many of the principles of aerodynamics. Gliders posed the opposite problem from that presented by hot-air balloons: gliders were difficult to launch, but could be steered and maneuvered fairly readily once airborne. Thus began a theme that recurs several times in the history of human flight: experiments with gliders teach aviators how to *fly*—how to maneuver, steer, bank, and turn in the air—but not how to take off. In effect, these glider pilots, including the Wright brothers, set themselves the task of solving the second problem, maneuvering, before they had successfully conquered the first, getting up into the air. Takeoff in a glider posed a potent problem, addressed unsatisfactorily by using ramps, jumping off buildings or cliffs, and building strange slingshot-like arrangements that give new meaning to the imaginary devices of Rube Goldberg. Once airborne, practicing gliding enabled pilots to perfect their flying skills and gain control of steering and turning, even though takeoff and landing were trying and often dangerous.

Otto Lilienthal, a German mechanical engineer, manufactured a number of delightful machines that clearly owed their inspiration to natural fliers in the world around him. In 1868, he struggled with wing-flapping devices, known as ornithopters, but by the 1890s he was experimenting with lovely bat-like biplanes and monoplane gliders. They were beautiful if not very practical. The largest and most successful of these prototypical hang gliders had a wingspread of twenty-five feet; it was constructed of muslin stretched over a willow frame. Lilienthal used hilltops and earthworks to create natural updrafts suitable for takeoff and attained over 1,000 feet in a single glide (versus the Wright brothers' eventual 120 feet). His main concern was not to solve the takeoff problem in any truly effective way but to learn how to maneuver in the air. Initially, he designed his gliders for maximum stability of structure, the idea being that he would influence the glider's flight path by shifting his body weight rather than by moving parts of the machine. In fact, none of the parts of his early machines was movable. Weight-shifting soon proved impractical and hard to control, so his later designs included movable elevators or flaps pulled by wires. On August 9, 1896, Lilienthal crashed as he was experimenting with a glider with a movable elevator. A gust of wind thrust his wings up in front; although he shifted his weight forward to try to get the nose down, the glider simply sideslipped and plummeted to the ground. Lilienthal died of a broken spine the next day, his last recorded words being "Sacrifices must be made." They are carved on his tombstone.

The legend is that the tragic end to Lilienthal's career captured the imagination of the Wright brothers and kept the quest for flight alive. In 1896, Orville and Wilbur were bachelor brothers, twenty-five and twenty-nine respectively. They lived together in Dayton, Ohio, then and for the rest of their lives. Together they ran a number of businesses, including printing and publishing various newspapers as well as designing, manufacturing, and selling bicycles. They were mechanically minded, intelligent, and curious, dabbling in photography and carpentry as well. The horseless carriage had first been seen in Dayton that year, and the brothers, like many others, felt that a revolution in transport was at hand. Late that summer, Orville was stricken with typhoid fever, a life-threatening disease. Looked after by his brother, his sister Katherine, and a trained nurse, Orville recovered after six long weeks of fever and misery. The story, often told by the brothers themselves, is that Wilbur read of Otto Lilienthal's death in the newspapers as he sat faithfully in his delirious brother's room. If so, Wilbur was catching up on some rather old newspapers, for Lilienthal died on August 10 and Orville did not fall ill until the last week of August. But, as the story goes, Wilbur waited until Orville was fully recovered to break the news of the tragic death because they had regarded Lilienthal as their hero and had been following his career avidly for a year or two. Lilienthal was known as the "Flying Man," and his exploits were covered by an eager press that pandered to the growing number of "air-minded" readers, including the Wright brothers. The idea that humans would soon learn how to fly now began to absorb the attention of the Wright brothers, who were eager to read everything they could find on the subject.

They appreciated Lilienthal's fundamental insight: that it was essential to learn how to fly *before* solving the problem of power for takeoff. In their own experiments in years to come, they followed Lilienthal's model, successfully mastering the skills and principles of aerodynamics while their takeoff apparatus was still terribly crude and wind power was all that sustained their flights. The Wright brothers accumulated hours and hours of experience in gliders, practice that positioned the pair for successful flights once the gasoline engine was available. Their first flying machine was a biplane with a five-foot wingspan and a tail that could be moved up and down as the upper wing moved forward or back. It was designed as a sort of kite and not as a transportation device, but experimenting with it revealed significant problems, the solutions to which led to a number of significant advances. The Wright brothers soon perceived that they needed some means of lateral control, for they were at first unable to induce a turn or control the rocking of the wings. Birds do this by altering the angles of their wings to restore balance and modern airplanes do it by moving the ailerons on the trailing edges of the wings in opposite directions, one up and one down. The principle behind these mechanisms was unknown in 1899, however. How could they establish lateral control in a flying machine? Wilbur always claimed that they hit upon the notion of wing-warping or wing-twisting while watching a flock of pigeons, although Orville later denied this anecdote. Whatever the inspiration, they decided to arrange a set of wires so that they ran from a pair of sticks in the kite flier's hands through a set of pulleys to the wing tips. Since the wings were constructed of lightweight wood covered in cloth, the natural torsional properties of the wooden frame allowed it to twist without breaking the wires. Wing-warping mimicked the twisting motion of the feathers and the wing used by birds to maintain lateral control. The Wright brothers soon incorporated this mechanism into a glider capable of carrying a human; it had two skids attached to its underside as a landing apparatus. Early in 1903, when they had learned enough from gliding, they decided to attach an engine and a propeller to their machine. But they couldn't find a motor that weighed less than 180 pounds and still produced minimal vibrations, so the Wright brothers built it and a redesigned *Flyer* to which their new engine would be attached. Further experimentation led them to an adequate design for the propellers. With sufficient lift, wing-warping, and an engine to provide power, the Wright brothers were the first humans to achieve powered flight before the year was out. Although nearly everything about the original Wright *Flyer* had been changed, refined, or totally redesigned in the process, the brothers had successfully resolved the three major challenges of flight—takeoff, maneuvering, and landing—or had at least arrived at a flying machine and a flying technique that was not deadly. Many other early aviators were literally doomed to severe injury or death by their inability to steer, turn, control, and land an airplane.

Although I know that human and avian flight differ in meaningful ways, tracing the history of the development of viable flying machines told me a lot about the evolution of avian flight. There is a substantial overlap in the difficulties of avian and human flight, strictures imposed largely by the conditions of moving a relatively fragile living being through the air in defiance of the force of gravity. These primary conditions of flight impose certain unalterable prerequisites no matter who the would-be flier is. One essential need for flight is lightness, a need solved somewhat differently in these two cases (hollow bones and feathers for birds, split-wood frames and lightweight fabric for humans), yet with an overriding similarity. Power for takeoff and landing is another problem, solved in birds by massive enlargement of the flight musculature and clever rearrangements and evolutionary modifications of the skeletal structure. For humans, it has proven easier to solve this problem through technology—developing a lightweight but powerful engine—rather than waiting for natural selection to act. A final and perhaps very telling point is that the success of the Wright brothers, where so many before them had failed, lay in their conviction that they had to master the intricacies of gliding flight—learning to steer, maneuver, turn, bank, move higher or lower, and circle—before they attempted powered flight. The consequence of the inverse strategy, using powered flight *before* learning these skills, is told in the long litany of dead aviators in the early twentieth century: those who flew fast enough but not well. Had we a better fossil record of the evolution of flight, it might tell the same story through a record of failed flying creatures with powerful muscles but limited ability to steer.

Chapter 3.
Flight Plan

———

Avian and human flight can also be compared from a perspective that emphasizes the dynamic and behavioral components. I remember reading a wonderful passage in *Out of Africa,* in which Karen Blixen recalls an incident that occurred on the boat on her initial voyage to Africa:

> One evening, as we were going to play cards, the English traveller told us about Mexico and of how a very old Spanish lady, who lived on a lonely farm in the mountains, when she heard of the arrival of a stranger, had sent for him and ordered him to give her the news of the world. "Well, men fly now, Madame," he said to her.
>
> "Yes, I have heard of that," said she, "and I have had many arguments with my priest about it. Now you can enlighten us, sir. Do men fly with their legs drawn up under them, like the sparrows, or stretched out behind them, like the storks?"

How a bird, or a flying man, uses its legs in flight or in landing is not the only issue of interest. As I tried to discover how such a complex and integrated system as the flight apparatus evolved, many specific questions about the dynamics of flight formed in my mind. For example, what precisely do the wings *do* in a wingbeat cycle? Which muscles propel the bones and what positions do they adopt? What is the role of the tail—or of the furcula, or wishbone? And, since the chest muscles that move the wings change the size of the chest cavity that houses the lungs, how does a bird manage to breathe and fly simultaneously?

Leonardo's conviction notwithstanding, simply watching and dissecting birds is not enough to discover the key to flight. Trying to build flying machines—a mechanic's version of trying to "evolve" flight—has clarified the basic problems involved in aerodynamics: taking off, steering, and landing. Still, reviewing the human struggle to build a flying machine couldn't tell me the particular ways in which birds solved these common aerodynamic problems. My aim was to try to retrace the natural process by which some reptilian ancestor evolved into a feathered and flying bird, so I needed more information and more detail. Learning exactly how a bird flies would, I hoped, provide the lens through which the fuzzy and imprecise fossil record of early birds and would-be birds could be interpreted. Historically, it was only when scientists began to perform experiments—some bizarre, some ingenious—that many of the previously unanticipated aspects of the flight plan of birds became evident.

One of the first and most famous experiments involving bird flight was performed by anatomist Max-Heinz Sy in the 1930s. He took a straightforward if faintly grotesque approach to determining the precise function of various flight muscles, especially the supracoracoideus, which performs the wing flip in modern birds. In Sy's day, there was considerable debate over whether or not the supracoracoideus also functioned during sustained level flight. Was wing flipping a special takeoff movement, or was it part of each and every upstroke? Sy addressed the problem directly: he cut the supracoracoideus tendon on a series of anesthetized crows and pigeons to see which functions were impaired. The supracoracoideus tendon is an easily located and distinctive structure, so the possibility of damaging other structures during surgery was minimal. This approach, while revealing, is only slightly more refined than cutting off one leg of a human to see if both are necessary for running. Once the birds had recovered from the surgery, Sy encouraged them to fly. He found they were no longer able to take off from a flat surface because they could no longer perform a wing flip: a predictable result. However, once the birds were aloft—if Sy tossed them into the air or allowed them to take off from a raised perch—they were perfectly capable of sustained flight that appeared to be normal. While it is important to think of the flight muscles as an integrated, functional unit, this experiment revealed a subtle division of labor among them. Sy concluded that the supracoracoideus acted only during takeoff and not during flight per se. But he had no way of determining whether the flight of his experimentally altered crows and pigeons was truly normal or not, nor could he know whether other muscles were working harder to compensate for the loss of the supracoracoideus. All he validly concluded from his crude experiment was that while the birds did not lose all ability to fly when the supracoracoideus was disabled, they did become incapable of takeoff from the ground.

Sy's study provided the first glimpse of which muscles were used and how and when they were used during flight, but the matter was far from settled. Studies of avian anatomy and analysis of conventional films of birds in flight let later researchers deduce what muscles must be acting when, but they had still no direct evidence for their conclusions. In the 1980s, when far more sophisticated techniques became available, the true modern heirs of Leonardo emerged. I turned to an old friend, Farish Jenkins, Jr., a Harvard anatomist who, with his collaborators G. E. "Ted" Goslow, Jr., now of Brown University, and Ken Dial, now of the University of Montana, had made some major discoveries about bird flight. Dial, especially, has continued this line of work and in recent years had enlarged the collaboration to include Stephen Gatesy of Wake Forest University as well. This is a loosely affiliated group, rather than an organized team, united by friendly ties among the researchers who share information, offer advice, and bounce ideas off one another regularly. Their work has transformed our understanding of bird flight.

The most significant advance lies in their technical wizardry, which far surpasses Sy's primitive techniques of dissecting dead birds or severing tendons of live ones. The first problem they wanted to solve was how the bones of the bird skeleton moved during flight. They decided to document the changing positions of the bones directly, using continuous radiographic films, or what is sometimes called cine X-ray. To do this, they trained some birds (European starlings in the initial work by Jenkins, Dial, and Goslow) to fly in a wind tunnel at speeds of nine to twenty meters per second, or about twenty to forty-five miles per hour. This was not as simple as it sounds.

They needed to get the birds to fly continuously at a particular point in the wind tunnel (like swimming in place in a lap pool), so that the radiographic camera could be mounted in a fixed position. They chose two views. In some sequences, the camera was placed above the bird's back (which gave them a dorsoventral view); in others, the camera was placed to one side (for a lateral view). The camera they used takes two hundred frames a second, but unlike conventional high-speed cameras, which record the movements of gross external structures such as the entire wing, the cine X-ray documented the movements of each bone throughout each wing flap. In their preliminary run-through, they found that it was extremely difficult to locate some of the particular skeletal features on each frame of the film because of the complexity of the images. Sometimes the images of the bones on one side of the bird became confused with those on the other side, because the X-rays can "see" all the way through the bird. The images effectively rendered the bird transparent and two-dimensional. To clarify the images, they decided to implant tiny, radio-opaque steel markers—each was smaller than a printed period on this page—at various anatomical points on the birds. They also inserted a short piece of very fine wire under the skin on the back, which would show up in each frame as a scale bar. The surgery was finicky and had to be performed under deep anesthesia, but the birds recovered quickly and seemed not to notice the tiny implants at all.

The team was particularly interested in understanding the roles of the furcula and the carina (the keel of the sternum, or breastbone) that are such distinctive features of birds. As any observant eater of a chicken breast can attest, there are two big muscles on a bird's chest. The external or most superficial layer of muscle on the bird's chest is the pectoralis, the same muscle that becomes so prominent on the chests of human bodybuilders. The avian pectoralis is even proportionately larger, for it is the major muscle that pulls the humerus (and thus the entire wing) downward in the downstroke. Deep to the pectoralis lies another separate, large muscle, the supracoracoideus. This muscle, as Sy established, is crucial in the wing flip. Logic dictates that the supracoracoideus should also work during the upstroke in normal birds because its primary action is to lift the humerus and lifting the humerus will automatically elevate the entire wing. Despite Sy's experiments, it seems strange and unlikely that such a large and functionally suitable muscle would not participate in the upstroke, and the cine X-ray films might resolve the problem. If the supracoracoideus does power the upstroke, then it and the pectoralis act reciprocally as a pair of antagonists, one muscle lifting the wing up and the other thrusting it downward. Because both wings beat in tandem, the movements of the right supracoracoideus and pectoralis are exactly coordinated with the movements of the left supracoracoideus and pectoralis.

These observations don't explain why the thoracic and shoulder bones of a bird are so unusual in shape. Other animals, including the presumed reptilian ancestor of birds, have two separate clavicles or collarbones and a long, flat sternum as compared to the bird's single furcula (wishbone) and its prominent carina, or keel, on the outside of the breastbone. Evolutionarily, the furcula was created by fusing the two clavicles into an odd, U-shaped structure that is characteristic of birds. Developing a carina on the sternum provided extra areas of bony attachment for the pectoral muscles as they grew larger and larger. Like the furcula, the carina is apparently of such pivotal importance for flight that every living, flighted bird has one, and no other creature on earth does. But the shape of these bones is not, strictly speaking, dictated by the action of flight, for bats fly and yet have neither a furcula nor a carina on their sternums. The secret must lie in something distinctive about how *birds* fly. Jenkins, Dial, and Goslow believed that documenting the precise movement of each of these bones would help them unravel this riddle. Because muscles are the only way an organism has to move its bones, knowing how the bones move during flight would tell the trio a great deal about muscle action as well.

After long days of preparation, the trio were able to take some incredible films of their starlings in flight. Their success was a technical tour-de-force and a tribute to their patience, skill, and persistence. Each flight generated a huge stack of films, each image of which needed to be painstakingly analyzed before they could understand bird flight. As a first step, they divided the wingbeat cycle into four phases: 1) the upstroke-downstroke transition; 2) the downstroke; 3) the downstroke-upstroke transition; and 4) the upstroke. While it is obvious that the upstroke and downstroke are important, the transitions between them are not so readily apparent. Yet during those transitions, the direction of movement and the effective shape of the wing as an airfoil are altered. The success or failure of an attempted flight depends directly on the bird's ability to make these adjustments rapidly and accurately. The transitions are also, logically, the points at which the pectoralis gives way to the supracoracoideus (or vice versa). Fine coordination and exquisite control over both muscles are essential for these transitions to be smooth and reliable. And, of course, other muscles work to alter the position of other bones in the wings, extending or flexing joints and thus adjusting the size and shape of the airfoil throughout the wingbeat cycle.

The drawings in Figure 15 are taken from the cineradiographic films generated in this study. They show the position of the wing during the transitions at mid-upstroke and mid-downstroke. The down-stroke generates the lift to keep the bird aloft as well as propelling it forward. The bird's wing is fully extended, to present a maximum surface area to the air. The wing pushes down and backward, thus moving the bird up and forward. The motion is analogous to swimming or rowing through the air. The reverse happens during the upstroke, or recovery. The wing is fairly tightly flexed during the upstroke, which diminishes the effective surface area of the wing and thus

Figure 15. Tracings taken from cineradiographic movies of a starling in flight show how the wing and bones move. From left to right, these images show: (1) the transition between the upstroke and the downstroke, (2) the middle of the downstroke, (3) the end of the downstroke, and (4) the middle of the upstroke.

prevents the recovery stroke from propelling the bird backward or downward. During the upstroke, the bird doesn't need to push against the air; it needs only to reposition its wing so as to be prepared for the next downstroke. Naturally, during the transition between upstroke and downstroke, the wings are partly flexed in an intermediate position. These findings were similar to those of earlier studies that had used conventional high-speed photography. What was not at all obvious before this clever study was that the furcula and the sternum also undergo a cycle of movements that are coordinated with the wing strokes.

The cineradiographic films show that the sternum actually pumps up and down throughout the wingbeat cycle. During the downstroke, the sternum moves backward (toward the tail) and upward; during the upstroke, it moves downward and forward (Figure 16). The movement of the bone is obviously produced by the contraction and relaxation of the flight muscles of the chest, which are anchored to the carina of the sternum. As the muscles contract to provide thrust in the downstroke, they pull the wings downward and forward; this is the main purpose of the muscular action. But this contraction also brings the sternum slightly upward and backward, too. Even though we think of the sternum as the bony anchor for these muscles and the humerus as the mobile element, the sternum is not rigidly fixed in place by any external mechanism, so it moves, too. Similarly, when the chest muscles relax and the back or dorsal muscles act to pull the wings upward and backward, the sternum inevitably moves in the opposite direction, forward and downward.

At the same time that the sternum is pumping up and down, the furcula is moving, too. It had long been thought that the furcula was primarily a mechanical strut or "spacer" keeping the shoulders in the appropriate position, as do clavicles on mammals. But living bone is somewhat flexible and elastic; the open U-shape of the furcula renders it vulnerable to distortion as muscles pull on it. As the cineradiographic films show clearly, the pattern of distortion of the furcula is important and cyclical (Figure 17). During the downstroke, the two tips of the furcula are pulled outward by the chest muscles, making the furcula more V-shaped than U-shaped. In the upstroke, the furcula recoils, flexing inward and resuming its original configuration. The distortion in

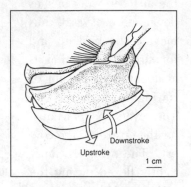

Figure 16. The sternum makes a pumping motion during the wingbeat cycle. It moves downward and forward during the upstroke and upward and backward during the downstroke, changing the size of the thoracic cavity.

shape is significant. In the resting or upstroke position, the tips of a starling's furcula are about 12.3 millimeters (slightly less than half an inch) apart. In the downstroke, these tips spread apart by more than 50 percent to encompass an average span of 18.7 millimeters. In other words, the right and left pectoralis muscles act just like a pair of children pulling on a wishbone to make a wish. The biggest difference is that the muscles, unlike the children, stop short of snapping the furcula. And, of course, the muscles are acting on living, flexible bone, and children play with dead bone whose elastic properties have been degraded through cooking.

Putting the movement of the two bones together, Jenkins and his colleagues realized that the shape of the whole thorax was altering throughout the wingbeat cycle. During downstroke, the thorax is wide and shallow; the furcular tips spread apart and the sternum comes up toward the backbone. During upstroke, the thorax becomes narrow and deep. The furcula releases the stored energy, recoiling to its original position like a stretched spring that has been freed from tension. This spring action of the furcula is coordinated with the lowering and forward displacement of the sternum.

What effect does this thoracic change have on the bird? A major one. The lungs occupy most of the space in the thorax. In a mammal, the expansion and contraction of the thorax force the lungs to expand and contract, sucking in or expelling air. But the respiratory tract of birds is more complicated than that of mammals. The lungs themselves are fairly rigid and do not expand or contract much during respiration. How-ever, the lungs are supplemented by a series of air sacs: an interclavicular air sac that lies just behind the arms of the furcula; a pair of cervical air sacs that lie just above the anterior part of the lungs; two pairs of anterior and posterior tho-racic air sacs that lie beneath the lungs; and a pair of abdominal air sacs that lie behind the lungs. Small air sacs also in-vade the bones and some of the muscles of the thorax. Air moves in through the bird's nostrils, down the trachea, and through this complex system in a one-

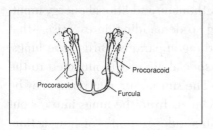

Figure 17. The furcula or wishbone seems to act as a spring during the wingbeat cy-cle. During the downstroke, the ends of the furcula are spread apart as the pectoral muscles pull the wings downward and for-ward. The furcula recoils to its original po-sition when the pectoral muscles relax, enhancing the upstroke and bringing the wings back to their upright position.

way flow, passing from the posterior air sacs into the lungs, and then from the lungs into the anterior air sacs and out through the trachea. (In contrast, in mammals, air passes into the lungs where gas exchange occurs and then back out of the lungs.) While avian lungs perform the task of gas exchange, remov-ing carbon dioxide from the blood and replacing it with oxygen, the air sacs cannot assist in this function because they lack capillaries. Air sacs are thought to function partly as holding chambers and partly as cooling mecha-nisms, since heat from the muscles can pass directly into the air in the air sacs and be expelled.

Although the avian respiratory system is complex, everything moves in rhythm with the wingbeats. Birds inhale on the upstroke, when the thorax expands, and exhale on the downstroke, when the thorax contracts. Yet this coupling is not an obligate arrangement because the lungs are remaining roughly constant in volume, saving birds from the tyranny of wings-up/inhale, wings-down/exhale. Birds can choose to breathe in a faster rhythm if they get tired or hot, the way a human runner can; during flight, starlings show an average respiratory rate of three cycles per second, while their wings beat at about twelve to fourteen cycles per second. The physical action of flying works like a bellows, moving muscles and bones to pump air in and out of the respiratory system with every flap. But rather than devote extra muscular energy to breathing and pumping, birds take advantage of the natural flexibility of the bony tissue of the furcula. The furcular tips are stretched apart as a secondary consequence of the flapping motion; the elastic recoil of the bony tissue recaptures the stored energy and completes the pumping action "for free"—that is, without any additional energy input.

Oxygenation is so crucial—and flying so demanding an exercise—that birds have further enhanced their breathing apparatus. Birds have lungs, like any vertebrate, but they also have special air sacs (connected to the lungs) that lie inside their hollow bones. The sternal/furcular pump may be a secondary mechanism that helps to move air from the lungs into (or out of) the air sacs. This mechanism operates independently of lung action, thus providing extra oxygenation during flight. Thus the air sacs enhance respiration. They also perform an additional physiological function that is key to avian flight, for they are part of the thermoregulatory system. In other words, the avian respiratory system not only provides oxygen to fuel rigorous activity, it also acts in an ingenious way to prevent overheating.

Heating and cooling are physiological imperatives for warm-blooded organisms like birds or mammals. The essence of being warm-blooded is that you maintain an approximately constant body temperature regardless of the weather outside. Tricky and energetically expensive as this maintenance may be, it offers the immense benefit of enabling the body to engage in demanding exercise at any time of day or night, not just when the air temperature is warm. Cold-blooded animals, such as lizards or crocodiles, are notoriously sluggish when it is cool and their body temperatures fall; they are far more dependent than warm-blooded animals upon their environments and on behavioral mechanisms, such as basking or seeking the shade, to control their body temperature. In most organisms, whether warm- or cold-blooded, the lungs and the bloodstream are the primary cooling mechanisms. The bloodstream absorbs heat from the muscles and carries it to the lungs or skin where it is dissipated by convection. But birds have improved upon this mechanism with their air sacs, which give them a greater surface area through which to dissipate heat. In fact, the air sac within the furcula branches into a network of tiny air cavities that spread out into the tissue of the pectoralis and supracoracoideus muscles themselves. This expanded respiratory system boosts the efficiency of the bird's cooling mechanism considerably. Much of the heat that needs to be dissipated is generated directly by muscle contractions, and the flight muscles—contracting and relaxing with every flap—put out a lot of heat. Most animals rid themselves of muscular heat simply by carrying it away in the bloodstream and dissipating it in some other part of the body where the blood vessels are close to the surface. But birds have elaborate, air-filled "cooling coils" layered directly into the big flight muscles, so the heat passes directly from the muscle to the air sacs and out of the body with every breath. This fact offers a new perspective on the anatomy of the sternum in birds.

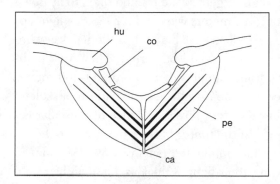

Figure 18. The pectoralis muscle (pe) of a bird is infiltrated with air channels (shown in bold) that help dissipate the heat generated during flying. The carina (ca) or keel of the sternum acts as a bony strut that prevents these air channels from collapsing when the pectoralis muscle is contracted, bringing the humerus (hu) downward. The coracoid (co) acts as a spacer. A bat has an unkeeled sternum and a different cooling system.

"This is the reason for the bony carina on the sternum of flying birds," says Colin Pennycuick, an expert on bird flight formerly at the University of Miami. "It prevents the pectoralis muscles from collapsing their internal air cavities when the muscles contract." Remember that the carina protrudes from the sternum at right angles, like the stem of a T. Remember, too, that the pectoralis and supracoracoideus muscles are anchored into the carina. This means that the muscles cannot compress and collapse the air cavities during contraction because the carina serves as a bony strut supporting them (Figure 18). Why do mammals do things differently? Because they have to, Pennycuick adds. "Bats use a different method of cooling, in which heat is removed from the muscles by the blood and carried to the wings, where it is disposed of by convection. Having no air cavities in their muscles, bats do not need a carina." And, I realized, bats have unfurred, skin wings from which heat evaporates readily while birds do not. Feathers are excellent insulation for keeping heat in or cold out; witness the immense popularity of goose-down comforters and jackets in cold climates. Since feathered wings are naturally insulated against both heat loss and heat gain, the cooling options favored by mammals were unsuitable for birds once they had evolved feathers, which was at least 150 million years ago. The special features of the furcula and carina are a clever evolutionary response to the need to cool and heat a highly active animal that is already so well-insulated.

The next step in the intricate tango of physiology and anatomy was documented by Dial, Gatesy, and another collaborator, Dona Boggs of the University of Montana. The cine X-ray studies had established the basic pattern of skeletal movements and breathing, so Dial and his colleagues focused on documenting the precise patterns of muscular activity. They used wild-caught, common birds (rock doves or pigeons) and trained them to fly down a hallway that was 50 meters long, 3.1 meters wide and 2.7 meters high (approximately 164 feet by 10 feet by 9 feet). At the end of the hallway were landing platforms the height of which could be varied. They took high-speed films of the birds' flights, using cameras that record 200–400 frames per second. They no longer needed cine X-ray, since the pattern of skeletal movements was known, but they did want direct evidence of muscular activity that could be calibrated against the timing of the wingbeats. So the researchers anesthetized the birds and implanted pairs of tiny electrodes in their muscles, using devices with wires about as thick as a human hair (100 microns) that could sense the electrical activity generated by muscular action. The electrodes were attached to an amplification and recording device by a very lightweight cable. These miniature devices recorded the timing and intensity of electrical impulses generated by as many as six different muscles at a time. In all, the research team documented the actions of seventeen different flight and wing muscles.

They defined different aspects of flight quantitatively, investigating the muscular activity involved in each. For their purposes, *takeoff* was the first five to eight wingbeats after the bird left the ground. In *level flapping flight,* the bird had to cover at least 30 meters of the hallway (about 98 feet). *Landing* was considered to be the last five wingbeats at the end of a level flapping flight. To study *ascending* and *descending flight,* the experimenters either allowed pigeons to take off from the floor, flying down the hallway to land on a perch 2.5 meters (a little over 8 feet) above the floor, or they released the pigeons by hand at a height of 2.5 meters almost directly above the landing platform and then allowed them to fly and land. They repeated the experimental flights until they had collected information on five wing-beat cycles for each muscle for each aspect of flight. The result was a complicated and enormous database of various types of information.

While some muscles functioned exactly as predicted—pectoralis in the downstroke and supracoracoideus in both takeoff *and* upstroke, for example—there were some surprises, too. For example, the data showed that all muscles were active in all aspects of flight; what varied was the intensity of their activity. Thus, to produce the downstroke, the pectoralis was very active, but its antagonist, the supracoracoideus muscle, was also active in order to modulate and control the movement smoothly. In turn, the supracoracoideus was the primary muscle of the upstroke and takeoff, but also functioned during vertical and descending flight. This muscle was least active during level flight, which explains why Sy couldn't see anything abnormal about the level flight of the birds whose supracoracoideuses had been severed. Apparently, although birds normally use the supracoracoideus during level flight, Sy's birds were able to compensate for its loss reasonably well.

Because the high-speed films were calibrated against the electromyographic readings, Dial and his colleagues discovered some important facts. Each individual bird seemed to have its own preferred wingbeat frequency, in the same way that humans of different sizes and shapes have different characteristic walking speeds that seem most comfortable to them. Whatever absolute frequency a particular bird preferred, it increased the frequency of its wingbeats significantly during takeoff and vertical ascending flight. In other words, birds flapped faster (more flaps per minute) to climb upward. Similarly, birds also flapped harder, increasing the amplitude (or size from top to bottom) of the wingbeat in some tasks. Once again, takeoff and vertical ascending flight demand the greatest wing excursion. When I heard that wing amplitude increased during takeoff, I realized that this explained a puzzling experience that I have had. Birds generally fly nearly soundlessly, but when a bird I haven't seen is lying in the grass and takes off unexpectedly from beneath my feet, I am startled both by the movement and by a loud clapping noise. Why should a bird fly noisily sometimes and silently at others? The answer is that the extreme amplitude of the wingbeat that is required for takeoff causes the wings to collide with each other at both the top and the bottom of the wingbeat. The high-speed films taken by Dial, Boggs, and Gatesy show the wings clapping together at the transition between downstroke and upstroke and again at the transition between upstroke and downstroke. Obviously, clapping the wings together is energetically wasteful, but less potentially harmful than being stepped upon or caught by a fierce predator, such as a human. In contrast, the wings do not collide at all in normal flight because the wingbeat amplitude is smaller.

Finally, in another, similar study, Dial was able to demonstrate a remarkable fact: *birds don't use their wing muscles to fly* under ordinary circumstances. Wing muscles are not the same thing as flight muscles. What Dial means by "wing muscles" are those that are completely confined to the wing; they originate from and insert into the bones of the wing itself and do not attach to bones outside the wing such as the carina or the furcula. It seems paradoxical that these intrinsic wing muscles don't work during flight because biomechanically they are positioned to extend or fold the wings, adjusting their total surface area. Yet the electromyographic data, derived from experiments repeated many times on different birds, show that the intrinsic wing muscles are inactive during sustained, level flight. Only the large flight muscles are used to produce the flapping movements of the wings. The flexion and extension of the wings—the folding and opening that change the effective area of the wing during different parts of the wingbeat cycle—are performed not by muscular activity but via the mechanical linkage mechanism created by the shape and articulation of the humerus and radius, described earlier. If wings are for flying and wing muscles move wings, how can a bird fly without using its wing muscles? At one level, the answer is simple: wing muscles aren't responsible either for flapping wings or for extending and flexing wings.

Then why do birds have wing muscles at all? This is a good question. One of the most valid axioms of biomechanics predicts that flying organisms should have few if any intrinsic wing muscles. In general, an animal can increase its locomotor speed over the course of evolution by diminishing the weight of the muscles in the distal part of its limbs (meaning the segments farthest from the body). Lighter and less bulky distal segments can be moved more swiftly; elongating distal segments increases stride length and is another common feature of swift animals. Thus in antelopes, horses, and cheetahs, for example, most of the muscle mass is found high up in the limb, near the hip or shoulder. Only the tendons, and not the muscle bodies themselves, are found in their lower legs and feet, so the lower segments of the limbs are long and slender. If birds followed the same biomechanical principles, then logically they ought to have enlarged their muscles and moved them up onto the chest and shoulder area, which has occurred, while minimizing or eliminating the wing muscles altogether, which has not. Instead, wing muscles have been retained. Why? Dial sees the evolutionary retention of the intrinsic wing muscles as a backup or an emergency system, "a consequence of the fact that those muscles are needed for modification of the shape of the wing during periods of nonsteady flight." The wing muscles aren't required to produce the flexion and extension of the wing during every flap, because the shoulder muscles and mechanical linkage system are sufficient to do that without using muscular energy. But when extra maneuvering is called for— perhaps to escape a stooping hawk or to avoid colliding with branches in gusty, windy conditions—birds can adjust their wing shapes more precisely

no matter where in the wingbeat cycle they are. It is like having an auxiliary steering mechanism, one that is too expensive to be used under ordinary conditions but can save your life in case of emergency.

One of the most interesting consequences of Dial's hypothesis is that it solves the long-standing mystery of how birds perform long-distance migratory flights. Like many people, I have been puzzled by the phenomenon, wondering how a tiny bird can fuel flights of many thousands of miles. If a bird gorges in advance, it only increases the weight it must support in the air; if it stops frequently to feed, there is less time each day to cover distance. And some of these flights include long spans over water, where landing, feeding, and resting are either dif-

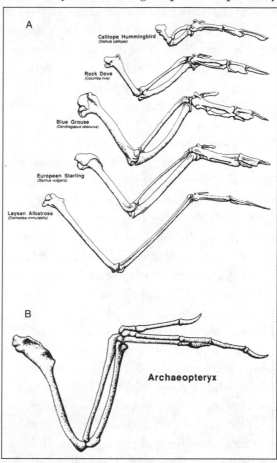

Figure 19. The shape and proportion of the wing bones of various birds (A) reflect the type of flight in which they specialize. The hummingbird, the specialist in hovering, nonsteady flight, has robust bones and a radius and ulna that bow away from each other to allow room for the attachment of strong forearm muscles. In contrast, the albatross, the consummate practitioner of long-distance, steady soaring, has elongated, straight, and fairly slender bones, with very little space between the radius and ulna. *Archaeopteryx* (B) shows only modest bowing of the radius and ulna. Not to scale.

ficult or impossible. But Dial's ideas about intrinsic wing muscles imply that when a bird undertakes a prolonged bout of level flapping, it can almost completely shut down all muscular activity within the wing. He estimates that level flapping probably occurs at a fraction of the metabolic cost of unsteady, wing-muscle-assisted flight. Another energy-saving mechanism is soaring, in which a bird spreads its wings and cruises on prevailing air currents and thermals. It is an important energy-saving strategy most often practiced by larger birds with long, narrow wings; they use soaring on both a daily basis and during migration. Smaller birds generally opt for flapping flight, which minimizes the time expended during migration, rather than undertaking prolonged bouts of slow soaring.

Dial also shows that the shape of the wing bones (primarily the radius and ulna of the forearm), and thus the size and shape of the wing muscles, differ according to the type of flight in which a species specializes (Figure 19). Hummingbirds, for example, are the consummate fliers, capable of hovering while they drink nectar, engaging in stunning aerial "dogfights" with rivals, and maneuvering in and out of twisting vegetation with ease and agility. They and other birds with similar capabilities have a radius and ulna that are relatively robust and stoutly built, by avian standards, and bow away from each other. A stout sheet of connective tissue stretches between the radius and ulna, whether they are bowed or closely spaced. When the bones are separated more widely from each other, the area of this connective tissue sheet is increased. This in turn provides an increased area of insertion for many of the intrinsic wing muscles, which attach directly to this sheet. Thus, if having larger, stronger intrinsic wing muscles is advantageous—allowing a bird to exploit an important resource not available to others, for example—then natural selection will favor the evolution of a bowed radius and ulna and an enlarged area of attachment for the wing muscles.

The opposite specialization from that of the hummingbird is seen in the albatross, an archetypal gliding bird with long wings adapted for long gliding flights over the ocean. The radius and ulna of an albatross, or another species with similar habits, are quite different from those of the hummingbird. The radius and ulna of the albatross are elongated, slender, closely opposed, and very straight; the mass of the wing musculature has been reduced in such species. Typically, the albatross engages in slope soaring, using the thermals generated by the surface of the waves and flapping very little. The anatomical features of the wings of an albatross actually hinder its ability to alter the shape of its wings, producing the characteristically uncoordinated takeoff and landing that is associated with albatrosses. The evolutionary history of the albatross has sacrificed maneuverability and fine control of wing shape and movement in favor of an extreme adaptation to soaring, whereas the hummingbird's anatomy exemplifies the opposite compromise.

Birds are not simply fliers, however, which makes their circumstance even more remarkable. They actually have at least two distinct locomotor systems within one body—a forelimb system designed for flying and a hindlimb system designed for bipedal walking or hopping. Sometimes both the forelimb and hindlimb systems are further adapted to moving birds' bodies through water as well. In contrast, the reptilian ancestors of birds had only one main locomotor system, a four-legged means of moving about that is sometimes land-based and sometimes aquatically adapted. Gatesy and Dial have proposed a special term, *locomotor module*, to designate an anatomical region that functions as a single unit during locomotion or propulsion of the body. Birds have two locomotor modules, while reptiles and most mammals have only one. Within a module, all anatomical parts must be integrated to make finely controlled movements possible; the skin, nerves, blood vessels, muscles, and bones must work together to produce a smooth and efficient movement. Indeed, the more costly the movement (in physiological and energetic terms), the more finely honed must be the anatomical system that supports it. Since flapping flight is a very expensive proposition, it follows that the locomotor module responsible for flight must be very well designed by the forces of natural selection.

Dial and Gatesy suggest an evolutionary scenario that would explain how one-module reptiles might have evolved into two-module birds. As a first evolutionary step, they propose, avian ancestors must have stopped using their forelimb in terrestrial locomotion, with two immediate consequences. First, the forelimb was freed from a supporting function, making it available to become the basis of a novel, secondary mode of locomotion like flying. Second, bipedalism (habitual two-footed locomotion) and all of its concomitant anatomical subtleties had to evolve. As any child learning to walk might attest, going from four supports during locomotion to two is a tricky change. It is not just a matter of standing up. The body must be entirely rebalanced and the patterns of movement reestablished. In bipedal walking, the weight of the body pivots over a single support while the other leg swings forward. Walking on four legs is a lot easier, since in most gaits at least two feet are on the ground at all times.

Like Dial and Gatesy, many researchers into bird origins believe that the ancestral reptile was already bipedal when it began to evolve toward the avian condition. They find it easier to imagine a bipedal dinosaur evolving its forearms into wings, which would leave its primary locomotor apparatus (the hindlimb) fully functional, than to think of a quadrupedal one simultaneously abandoning quadrupedalism for bipedalism plus aerial flight. The bipedal ancestor hypothesis, while favored strongly by logic, has little direct evidence from the fossil record to support or refute it. The strongest point is the repeated confusion, historically, between *Archaeopteryx* and the small bipedal dinosaur, *Compsognathus* (Figure 20). Obviously, these two species are very similar anatomically, the major difference being that one has wings and feathers and the other (as far as we know) doesn't. Unfortunately, *Compsognathus* lacks one key criterion for being *Archaeopteryx*'s ancestor: it is the wrong age—contemporaneous with, not older than, *Archaeopteryx*. Paleontologists who support the bipedal ancestor hypothesis are thus left in the uncomfortable position of asserting that something *like Compsognathus*—something as yet unknown—must have preceded and been ancestral to *Archaeopteryx*. This may be true, but balancing on a hypothetical ancestor is an even shakier proposition than balancing on one leg while the other moves forward.

Prior to bipedalism, the starting point for evolving flight must have been a more primitive, simpler, and less expensive form of locomotion: quadrupedalism. Practiced by most reptiles such as today's alligators, croco-

Figure 20. Artist and dinosaurologist Greg Paul thinks that the small theropod dinosaur *Compsognathus* may have tried to catch and eat *Archaeopteryx* at Solnhofen.

diles, or lizards, quadrupedalism calls for a single locomotor module that encompasses the torso, both sets of legs, and the tail. Most of the time, the legs are splayed out, with the feet planted on either side of the torso (in a position known as *ab*duction) rather than tucked firmly underneath the torso (a position which is, confusingly, called *ad*duction). However, reptiles can still lift their bodies up off the ground for a sprint by adducting their limbs. In crocodiles, fast running may be as swift as thirty miles per hour over a short distance. When sprinting, a crocodile holds its tail stiffly out behind its body, using it as a rudder for balance when turning rapidly. Crocodiles also walk, with the body held lower and the tail undulating only mildly. The slowest gait is what might be called the slither, in which the torso is actually dragged along the ground by the fully abducted legs. In the slither, the tail and torso undulate strongly, tracing a moving S-shape when seen from above.

The tail is capable of undulating because such animals have unusually shaped joints between the adjacent pairs of vertebrae in their torso and tail. Unlike those of most mammals, the crocodilian tail vertebrae are connected by ball-and-socket joints, which permit a maximum range of movements up, down, and side-to-side. The movements of the tail are perfectly coordinated with the leg and torso movements for two reasons. First, many of the muscles that move the leg and torso actually produce the movement in the tail as a secondary function. Second, there are long muscles and strong ligaments that run down either side of the entire vertebral column, from skull to tail tip, ensuring that the vertebral column acts as a single coordinated unit. A muscular pull or push in one area will be translated throughout the entire spinal column. The true value of the lateral undulation of the muscular tail becomes apparent during swimming. A crocodile, an iguana, or any similar reptile swims with its arms and legs straight and flattened against the torso and tail. The limbs are useless for propulsion in this position and the body is propelled solely by the lateral undulations of the body and tail, in a motion mimicking that of the (limbless) snake. This is a perfect example of a single, beautifully integrated locomotor module that functions well both on land or in the water.

Bipedal lizards or dinosaurs use their tails differently from quadrupedal ones because of their different mode of locomotion. For them, a tail is primarily a counterbalancing weight that helps keep them from falling forward in the walk or run. The tail is therefore held stiffly extended behind the body, and its rigid posture is enhanced by elaborate bony articulations between pairs of vertebrae—not ball-and-socket joints—that limit movement severely. The strong ligaments that run along the vertebral column, structures that act like rubber bands or bungee cords in alligators, actually become ossified or bony in bipedal dinosaurs and are sometimes preserved in fossil specimens. This means that the lateral movements and strongly curving tail postures so important in crocodile-type locomotion were anatomically impossible for bipedal dinosaurs. The primary plane of movement for the tail vertebrae of such dinosaurs was up and down; the tail tended to act as a stiff unit that was hinged at the base of the tail, where it connected to the pelvis.

Modern birds, of course, are built differently from either crocodiles or bipedal dinosaurs, and accordingly their tails have yet another way of functioning in locomotion. With the success of electromyography in figuring out the bone and muscle actions of the wings during flight, it was inevitable that Stephen Gatesy and Ken Dial would turn their attention to the activity of the tail during walking and flying. As before, they implanted miniature electrodes in pigeons under anesthesia. Activity in the pectoralis muscle was monitored to indicate when the wing was being used, while a hip muscle, the iliotrochantericus caudalis, was taken as an indicator of bipedal walking. Nine different tail muscles were monitored, too, and the experiments were repeated until the pair had documented muscle activity during at least five cycles from two birds for walking, takeoff, slow, level, flapping flight, and landing. Because of the complexity of liftoff, the experimenters monitored that behavior over a larger number of trials.

The underlying skeleton of the tail of a bird is distinctive. The pelvic girdle is fused to the sacral vertebrae, to form a rigid, bony structure called the synsacrum. Moving toward the tail, there is then a chain of five or six vertebrae ending in the pygostyle. The pygostyle is a small, blunt bone created by the evolutionary fusion of several vertebrae; into it are inserted the six pairs of main tail feathers, or retrices. What you see when you look at a bird's tail is the retrices, the movements of which are controlled by the tail muscles. Anatomically, the muscles of a bird's tail are relatively isolated from the adjacent legs and trunk or wings and are controlled by separate nerves. This observation might suggest that the tail is in itself a third locomotor module, separate from wings and tail, except that no distinct locomotor mode can be traced solely to the actions of the tail. Yet the tail operates semi-independently from either the legs or wings, while still being carefully coordinated with them in order to make the bird's overall movements useful.

The studies by Gatesy and Dial have produced a wonderfully clear understanding of how the tail acts under different locomotor circumstances (Figure 21). In walking, the bird uses its tail muscles relatively little, although the actions that occur are closely coordinated with the movement of each leg. The most striking changes occur when a bird takes off, moving from terrestrial to aerial locomotion. In a standing bird, the tail is usually furled—that is, the retrices are clumped together—producing a cross-sectional shape that is concave-down or tented. As the wings are raised to initiate liftoff, the tail is also raised. The tail feathers flare as the retrices spread out and arch to form a concave-up shape—a movement known as "flaps up" to pilots. As the first downstroke is made, the bird moves its tail downward, tucking it under slightly, and flares the remiges even more widely. This produces a fan-like shaped tail that is nearly verti-

cally oriented. As the bird accomplishes take-off, its tail stays low and wide, although the flaring is diminished slightly every time the wings pass the tail in the downstroke. Once slow, level, flapping flight is achieved, the tail is reoriented to be less vertical and more horizontal. After that, the tail muscles are almost inactive, showing only a single burst of activity during each wingbeat. Finally, the bird again uses its tail actively in landing, returning it to the vertical, flared position—a pilot's "flaps down"—to help with deceleration. All this describes nor-

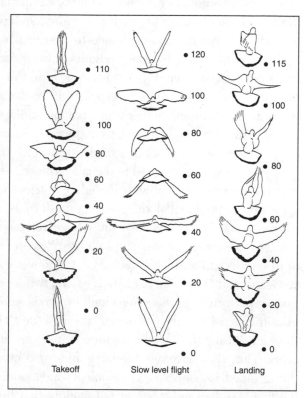

Figure 21. Stephen Gatesy and Ken Dial showed that a bird's tail moves in a unique pattern during flight. Each sequence begins at the bottom of the figure at time 0 and proceeds for 110–120 seconds. The black dots are fixed reference points. Birds evolved three independent locomotor modules: the hindlimbs and torso, the forelimbs, and the tail.

87

mal flight, but birds have to be prepared for gusty conditions, predators' attacks, and other unpredictable events. Although they haven't yet done the experiments, Gatesy and Dial guess that the tail muscles will be very active during unsteady flight. In the same way that the intrinsic wing muscles adjust the wing shape, the tail muscles probably alter the shape and angle of the tail to help stabilize the bird's flight under unsteady conditions. The tail thus has a role in both ordinary flight and as another element in the emergency maneuvering system.

The significance of documenting the action of various tail muscles and the way in which they are coordinated with wingbeat cycles is greater than might be immediately apparent. The unique avian locomotor system developed from a basically reptilian, quadrupedal pattern that was then modified by natural selection, in response to functional demands. With their up-and-down movements, birds use their tails more like bipedal dinosaurs than like crocodiles whose tail undulates side-to-side, making it more probable that birds descended from bipedal dinosaurs than crocodile-shaped reptiles. Another clue to the evolutionary transition—and thus to the probable origins of birds—comes from *Archaeopteryx* itself. *Archaeopteryx* has no pygostyle but instead sports a particularly long tail, including up to about twenty individual vertebrae. Ossified tendons and ligaments run the length of the tail, enhancing its ability to act as a single unit. Since there is no pygostyle, the pairs of retrices are attached to the sides of each vertebral body in the tail. *Archaeopteryx*'s long tail with spreading retrices makes a highly plausible transition from bipedal dinosaur to modern bird. Sometime after *Archaeopteryx*, the bird lineage must have reduced and modified its long string of feathered vertebrae into a pygostyle and rearranged the placement of the retrices to enable modern avian tail movements.

Gatesy and Dial think of birds as having two separate locomotor patterns that are performed by three, separately evolved locomotor modules. Birds hop or walk bipedally on the land and fly through the air (or, in some cases, under the water) using their hindlimbs for one purpose and their forelimbs for another. Thus they have one hindlimb-and-torso module, which is dominant during bipedal locomotion; a forelimb or wing module, which is dominant during flight; and an independent tail module that functions primarily during flight. This arrangement is unique to birds. Somehow birds evolved from the ancestral, single-module condition to have not one but two additional modules. Dial and Gatesy have predicted that the fossil record will eventually show that the locomotor modules split at two different points during the evolution of birds. First, activities of the forelimbs and tail became separated from those of the hindlimb, pelvis, and torso. This uncoupling meant that the muscles and nerves of the tail and forelimbs were freed to establish a novel and intimate affiliation with each other as the forelimb became adapted for flight. Meantime, the hindlimb continued to evolve for a specialized form of terrestrial locomotion in which two legs alternated rather than four. Thus logic, anatomy, and paleontology all support the same deduced sequence of evolutionary changes between reptiles and modern birds: bipedalism first; wings second; tail third. It is a brilliant hypothesis that can be tested as more fossil evidence becomes available to reinforce or, less probably, to refute their ideas.

Experimental studies of flight greatly improved my understanding of the end point of the evolutionary sequence, modern bird flight. But I also needed to understand the starting point, the reptile's walk. Since ancient reptiles cannot yet be re-created in the laboratory (*Jurassic Park* notwithstanding), scientists must be content with using modern reptilian locomotion as a model for the behaviors of ancient reptiles—a presumption that carries some risk of inaccuracy, though not a great one. Documenting, analyzing, and understanding different locomotor patterns are complex and demanding tasks, but ones in which surprising strides have been made. And much of the evidence bears directly on the pressing question: who was the ancestor of the first bird?

Chapter 4.

Nesting Sites

The ancestors of birds were reptilian and older than *Archaeopteryx:* on this everyone agrees. But there is widespread disagreement about *which* reptilian group was the ancestor of all later birds. And before searching for the nest in which birds were hatched, I had to figure out how to recognize a bird when I encountered one in the fossil record.

"Bird" is a mental category, but I don't know how humans form categories and classify objects in their mind. Is there an ideal, archetypal bird, to which we mentally compare all candidates for bird-hood? Of what does such an ideal consist? It is surely more than a simple list of attributes—birds have beaks, feathers, and fly—because we have no difficulty recognizing flightless birds as birds, nor does a plucked chicken cease to be a bird. Beyond those attributes lies a quality that we might call "birdiness" that even flightless, featherless birds partake of. If there is a quality of birdiness—a term that sounds suspiciously reminiscent of the Montgolfiers' attribution of "levity" to smoke—then how do we know of what that quality consists? Is birdiness one of the myriad cultural conventions that we learn from our elders as we grow up, along with how to speak and what names to assign to objects?

I believe our parochial or individual definitions of birdiness are learned. By the time we are a few years old, most of us know what a bird is or is not.

We can confidently place creatures we have never before seen into one of the two categories, bird or nonbird, and we may even try to assign a new creature to a specific category or type of bird. This is why colonists or explorers in strange places so blithely apply the names of familiar species to new ones they encounter. For example, the American robin was named by English immigrants who placed this new species into a preexisting mental category, based on the robin of their native land, which superficially resembles the American bird. I don't believe that these immigrants thought the New World bird was the same as the one they had left behind, but the new bird was more like a robin than anything else; sometimes, too, nostalgia for the world left behind has moved people to assign old names to new species. This common phenomenon of name transference is in part a traveler's attempt to render the strange familiar. Knowing the name of something or someone is a means of taming the wild, of acquiring power over the unknown.

Asked "What is a bird?," most people generally resort to one of two types of answers. The first can be thought of as pointing, verbally or literally: giving examples of the types of specific items that are found in the category "bird." Many respondents say something like this: "Well, a chickadee is a bird, and so is a parrot, and so is an ostrich or a penguin. Oh, and hummingbirds are birds, too, and vultures . . ." This is a tactic that relies on the listener's ability to extrapolate the essential attributes of a category or class of things (birds) from a review of some of its members. However, not everyone is equally observant or equally talented at making those judgments, nor will everyone be familiar with each particular example.

Another approach, sometimes combined with the first, is to try to list the defining attributes of the category rather than its members. Such an answer might be: "Birds are feathered bipeds. They fly—well, most of them fly or else their ancestors *used* to fly. They have wings and a feathered tail; they have beaks and lay eggs in nests; they sing and many types of birds migrate seasonally . . ." The problems with this descriptive response are also immediately apparent. First, the attributes are not sufficiently exclusive, because many of them are found in nonbirds. Turtles, for example, have beaks and lay eggs in nests and sing (or groan) after a fashion, but they are not birds. Bats have wings and fly, and so did pterodactyls, but they are not birds either. Even alligators lay eggs in nests, sing (or groan), migrate, and fail nonetheless to be birds. Second, the attributes are not sufficiently inclusive. For example, ostriches and penguins *are* birds but don't fly (except under

91

water in the case of penguins), don't have proper wings, and ostriches hardly sing and penguins only barely have a feathered tail. All birds lay eggs but they don't all make nests, unless "nest" is so broadly defined as to include a shallow depression on the ground. Some birds don't even have what could fairly be called real feathers.

The problem became acute when I tried to include fossil birds in the analysis. Starting with *Archaeopteryx* at about 150 million years ago, the fossil record documents an enormous adaptive radiation of archaic birds that disappeared and was replaced by more modern birds starting at about 65 million years ago. The birds that evolved during this early radiation are collectively called Mesozoic birds, for the geologic era in which they lived, which encompasses more than half of the known evolutionary history of birds. Within the Mesozoic era, most of these early birds are known from the latest period, the Cretaceous (144–64 million years ago); few are found in the Jurassic (213–144 million years ago) and there is no unequivocal evidence of birds in the earliest period, the Triassic (248–213 million years ago). The extinction event that marks the loss of the Mesozoic birds also saw the demise of the nonavian dinosaurs; it is sometimes called the K-T event because it is the end of the Cretaceous period (symbolized by K) and the beginning of the Tertiary (T) period, the first part of the Cenozoic era in which we still live. No Mesozoic birds can be directly linked to any modern forms, although some show adaptations that are functionally similar to those of modern birds, which appeared in a huge adaptive radiation subsequent to the K-T event.

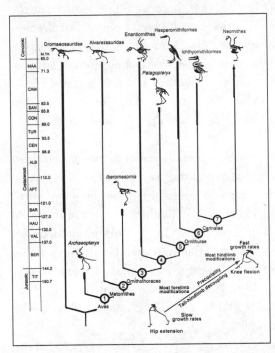

Figure 22. This cladogram of bird origins by Luis Chiappe suggests that *Archaeopteryx* was most closely related to the dromeosaurids among the theropod dinosaurs and was ancestral to all later birds. Bold vertical bars indicate the time from which fossil specimens are known; lighter bars indicate inferred relationships.

Larry Martin of the University of Kansas has argued that there is a basic two-way division among the Mesozoic birds from the beginning, although recent discoveries have complicated this simple picture and caused it to be challenged by other workers. In Martin's terms, one major group is the Ornithurae, most of which have aquatic adaptations and a modern sequence of fusion of the foot bones, while the other is the Sauriurae. The nonaquatic early birds of the Sauriurae are in turn divided into two smaller groups, the Archaeornithes (the "ancient birds") and the Enantiornithes (the "opposite birds"). The Enantiornithes are named "opposite birds" because their foot bones fuse in a pattern opposite to that seen in modern birds. *Archaeopteryx* is a representative of the Archaeornithes, but its exact position relative to later birds is debated. Martin views *Archaeopteryx* as ancestral only to the enantiornithines, while others, such as Luis Chiappe of the American Museum of Natural History, have constructed cladograms showing *Archaeopteryx* as ancestral to all later birds (Figure 22). Modern birds, which obviously have a modern sequence of foot bone fusion, fall into a subdivision of the Ornithurae called the Neornithes, or "new birds."

A series of important new fossil birds have been discovered in China, Spain, and Mongolia in the last few years, and these have complicated and confused the once-neat story of avian evolution after *Archaeopteryx*. The most spectacular new fossil species is *Confuciusornis*, an enantiornithine that has modern feathers arranged in a wing. The specimen comes from Liaoning, in northeastern China. Dated (with some uncertainty until more studies are completed) to between 121 and 142 million years, *Confuciusornis* may be only slightly younger than *Archaeopteryx*, but is very different from it. *Confuciusornis* has the oldest true, horny beak and the oldest pygostyle, whereas *Archaeopteryx* has teeth, no beak, and a long, reptilian, bony tail. Another specimen from the same site and age is called *Liaoningornis*, which is the oldest ornithurine. It is also feathered and, like *Confuciusornis* (and arguably *Archaeopteryx*), has feet adapted for perching. Its skull is primitive and toothy, but its flight apparatus is more advanced than that of *Confuciusornis*, with a bony keel on its sternum and a capacious and well-developed rib cage that hints at a modern avian style of respiration. Clearly all of the modern adaptations to flight did not develop at the same time.

This basic message, that evolution occurred in a mosaic fashion, is reiterated by the enantiornithines from Spain, dating to the early Cretaceous about 135 to 120 million years ago. These are small-sized species like *Noguerornis* and *Iberomesornis*. These both preserve evidence of a more modern shoulder and an enlarged wing. So does an unnamed nesting bird announced in 1997, whose complete skull has toothed jaws like *Archaeopteryx*. Also from Spain, and dated to 115 million years, is a goldfinch-sized bird, *Eoalulavis hoyasi*. It is a wonderful specimen. All of its feathers are preserved, as well as its last meal: a crustacean, perhaps a small crab, whose remains are visible in the bird's belly. *Eoalulavis* has a narrow sternum with a keel and a more modern shoulder joint than its predecessors, but retains primitive claws on its fingers like those on *Archaeopteryx*. Because of the unusually good preservation, you can also see that the small, first digit on its hand was feathered, making a bastard wing or alula. An alula lifts during slow flight, acting to redirect the air flow back to the surface of the wing and thus preventing stalling. It marks an important improvement in maneuverability among birds—although earlier birds that are less well-preserved may also have had an alula that has been rendered invisible by the circumstances of preservation.

Perhaps the most amazing archaic bird to be found in recent years is a species called *Mononykus,* known from two specimens found in Mongolia. This flightless bird is so unusual that the second sentence of the scientific paper announcing and naming the specimen—a paper that appeared in the normally conservative journal *Nature*—actually describes the specimen as "startling." It does not fit well within either the enantiornithines or the ornithurines. *Mononykus* is a large (turkey-sized), primitive bird with a long, bony tail and slender hindlimbs, and resembles *Archaeopteryx* in this and in details of its teeth, feet, and pelvis (Figure 23). It lacks a furcula—a normal circumstance for a flightless bird—but it does have a bony sternum with a ca-

Figure 23. *Mononykus olecranus* is a bizarre and controversial fossil from the Late Cretaceous of Mongolia. Although its forelimbs are adapted for digging and not flying, *Mononykus* is grouped with other early avians in a taxon called Alvarezsauridae because of avian features in its skull, sternum, feet, and pelvis.

rina. The carina, or keel, is not related to flight muscles, however, because the forelimb of *Mononykus* is extraordinarily modified and in no way resembles a wing. The forelimb is very short, robust, and ends in a single, large, and sturdy claw—"similar to that of digging animals," according to the team from the American Museum of Natural History and the Mongolian Museum of Natural History that found the specimens. They immediately add a cautionary note: "The short forelimb and long, gracile hindlimb are, paradoxically, incongruous with a burrowing habitus." Indeed, the specimen is so bizarre that a number of well-known figures in the *Archaeopteryx* debates—Larry Martin, Alan Feduccia, Peter Wellnhofer, and John Ostrom, to name a few—have expressed doubt whether it is a bird at all. A complete skull, found after the original announcement, shows additional avian features, but this has not turned the tide of opinion. In a recent review, Luis Chiappe (one of the team who found and described *Mononykus*) remarks,

> The placement of *Mononykus* outside Aves requires the less parsimonious conclusion that all these characters evolved independently in these two taxa [*Mononykus* and *Archaeopteryx*]. This fact has not avoided criticism in which *Mononykus* is *a priori* disregarded as a bird. . . . The criticism seems to stem from the conjecture that *Mononykus* does not fit the "stereotype" of a basal bird. . . . However, assumptions about evolutionary processes or adaptational scenarios . . . are misleading when identifying historical relationships [among fossil forms].

If indeed *Mononykus* represents an early flightless bird, it is probably not directly related to living flightless lineages.

Other Mesozoic ornithurines had extraordinary aquatic adaptations for swimming and diving. Some of these are almost as old as *Archaeopteryx*, such as some members of the group called Hesperornithiformes. These were diving birds, somewhat similar but not directly related to loons. They propelled themselves through the water with huge, webbed feet and, like *Archaeopteryx*, had teeth in their jaws. What they didn't have was wings. For example, *Hesperornis* is one of the best-known genera in this group, first found in Kansas in 1870 by the famous Yale paleontologist Othniel Charles Marsh. Its forelimb has only a vestigial humerus and no trace of a forearm, hand, or wing. Presumably the reduction of the forelimb to a tiny stub was part of streamlining the body for swimming and many of *Hesperornis*'s kin show similar adaptations. The feathers of a closely related species, *Parahesperornis*, are preserved and are what is called plumulaceous; the birds would have looked furry rather than feathered. At least three well-known species, *Hesperornis*, *Parahesperornis*, and *Baptornis*, show such extreme adaptations of the hindlimb for swimming that they were unable to bring their legs under their body to support weight. As Larry Martin, one of the world's foremost experts on Mesozoic birds, describes it, "This meant that on land the hesperornithiforms pushed themselves along on their stomachs probably with an undulating motion as seals do today." How and where they nested remains a puzzle, since their slow and awkward locomotion would have rendered them extremely vulnerable while on the ground. The Mesozoic was an era of strange birds doing strange things.

Only at the end of the Cretaceous, about 65 million years ago, do fossil birds directly related to modern birds, or Neornithes, first appear. The earliest specimens are related to modern birds with aquatic adaptations, like geese, ducks, flamingos, and wading shorebirds, although many of the ancient species seem to combine features of several modern forms rather than being clearly ancestral to any single one. There are also hints in the fossil record of the lineage leading to ratites—the group of large flightless birds including emus, ostriches, and rheas—in the form of a species called *Paleocursornis*, known from Romania.

Martin was one of the first to publish a paper endorsing the existence of the enantiornithines, whose existence as a group was suggested by Cyril Walker in 1981. That the existence of a major evolutionary event was unnoticed for so long, despite the discovery of many enantornithine fossil species from around the world, is a measure of how neglected the study of Mesozoic birds has been until recently. Martin is almost single-handedly responsible for the new appreciation of the importance of Mesozoic birds, which he began to study at a time when few paid any attention to them. With this solid grounding in the birds that followed *Archaeopteryx,* he entered the debate—and has also often acted as the pin in the balloon of dinosaur paleontologists working on the origin of birds.

A dark-haired, bearded, stocky man, Martin's genial appearance belies his love of a good argument and his unshakable belief in hard-nosed facts. He came to the controversy over the origins of *Archaeopteryx* and bird flight from the perspective of one who studied early avian evolution and functional anatomy: he arrived in the Jurassic by way of the Cretaceous, so to speak. In contrast, paleontologists like John Ostrom traveled within the Jurassic, arriving at avian evolution via the biology of dinosaurs that were roughly contemporaneous or even older. This difference in frame of reference has made a fundamental philosophical difference in Martin's and Ostrom's interpretations of the fossil material. Although it is something of an oversimplification, in debates over *Archaeopteryx* and bird flight, the ornithologists and paleo-ornithologists tend to stand on one side of the line and the dinosaurologists on the other.

The debate is analogous to a fight over kinship with the long-lost claimant to a family fortune. Ornithologists don't welcome dinosaurs into the family; they have nothing to gain by admitting these strangers and everything to lose. The diversity of anatomy and behavior of living birds (not to mention fossil birds) is enormous and provides fertile ground for endless studies, based on wonderful data: the proverbial bird in the hand. Consider a remark made by Ernst Mayr, the great ornithologist and evolutionary theorist at Harvard, to fellow ornithologist Alan Feduccia, after Mayr had been to Berlin: "The thing that strikes me about *Archaeopteryx*, Alan: that it's a *bird*." And birds—it goes without saying in any interchange between these two specialists—are not dinosaurs. Feduccia remarks that he wrote his first paper against the hot-blooded dinosaur theory in 1973 and its demise has been a long time coming. He does not—cannot—see any of the many purported detailed anatomical resemblances between dinosaurs and the birds he knows so well. Dinosaurs are weird and different beasts, from this perspective, who only contribute uncertainty and confusion.

But paleontologists who study dinosaurs—dinosaurologists—see things differently. They have enriched their lot immeasurably by adopting birds as living dinosaurs. Looking at dinosaurs with the knowledge of bird behavior and adaptations in mind has transformed our views of dinosaur biology. Two obvious examples can be found in the brilliant work of Jack Horner and John Ostrom. These men, among others, have totally revised scientific views of dinosaurs—and every step of this revision has tied them closer to birds. The revision has also reached the general public, thanks to their popular articles, Horner's engaging book *Digging Dinosaurs,* and Horner's contribution as adviser and role model to the highly successful films *Jurassic Park* and *The Lost World.*

Jack Horner is one of the world's great originals. He is a long, lanky man, and his open, straightforward manner of speaking reflects his Montana roots. He will talk with anyone, anywhere, about dinosaurs, and his tales of his discoveries are irresistibly entertaining. He is eager to learn about dinosaurs and will study anything, however arcane, that will help him understand them, but things that do not interest him do *not* interest him. Horner never bothered to finish college because the University of Montana wanted him to take some uninteresting (to him) courses in subjects like French and·

97

humanities. Armed with a high school diploma and a fervent love of dinosaurs, he tried to find a job in paleontology by writing "letters to all the natural history museums in the English-speaking world, twice, inquiring about work." In 1975, he was fortunate to find a job at Princeton University, as a preparator, the technician in the back room

Figure 24. Jack Horner and his colleagues found evidence of colonial nesting grounds of the duckbilled dinosaur *Maiasaurus* near Billings, Montana. This discovery was one of several that emphasize the behavioral similarities between dinosaurs and birds.

who cleans up the fossils, glues them together, and generally makes them "look interesting." His boss was Don Baird, a paleontologist who encouraged him to be a partner in research. Even though the dense population of the eastern United States made Horner feel cramped and crowded, he appreciated the wonderful museums full of dinosaurs. On his vacations, Horner could still prospect for dinosaurs in the open spaces of Montana, usually accompanied by a friend, Bob Makela. It was a viable compromise.

In the summer of 1978, he and Makela took a fortuitous detour from prospecting to stop in a rock shop in the tiny town of Bynum, Montana. They had heard the proprietor, Marion Brandvold, wanted help in identifying a few dinosaur fossils in the shop, and they were glad to oblige. Afterward, Brandvold asked if they would also look at some bones she had in the house. What she showed them, spread out on a card table, were numerous small bones of at least four baby duckbilled dinosaurs. It was like walking into a junk shop and being shown the missing crown jewels of the last czar of Russia.

The material was unprecedented, extraordinary, for two reasons. First, duckbills were Horner's favorite dinosaur and here was a treasure trove of them. Second, there were practically no known baby dinosaurs of any species in 1975. Mrs. Brandvold was happy to show them where the babies had come from, at a place near Billings. Within a few days, she had led them to the site and they had obtained permission to dig. In short order they found a clearly delineated dinosaur *nest,* a hollow deliberately scooped out by the mother as a place to lay her eggs (Figure 24). In all, they collected the bones of fifteen three-foot-long tiny dinosaurs and lots of eggshell fragments, as well as the geological evidence of the nest itself. A few dinosaur nests had been found previously, most notably in Mongolia in the 1920s by an expedition from the American Museum of Natural History. But Horner and Makela soon realized an even more amazing fact: all the babies were the same size. They had found a set of hatchlings from a single clutch, still in their nest along with the trampled bits of their broken eggshells. But because the first set of teeth in their tiny, three-inch-long jaws were worn, Horner knew they had been eating for some time. This meant they simply couldn't be newborns. Horner explains the stunning implications of this fact:

> If this was true, the next step in interpreting the fossils was to explain how they could have hatched together, stayed together, eating all the while, and then died in the nest. The most likely explanation is that they had never left the nest, and that one or both parents cared for them, bringing them food. It's conceivable that they wandered around in a group, feeding outside the nest and then returning there to rest, but I find this hard to believe. For one thing, this would have meant 3-foot dinosaurs walking among their 30-foot parents. That's a dangerous way to start life. I think they would have been safer in the nest.
>
> And yet if they were cold-blooded, like modern reptiles, it would have taken them a long time to grow . . . perhaps almost a year. . . . Living reptiles just don't do their growing in the nest; they get out as soon as they hatch. And warm-blooded creatures such as birds, which do quite a bit while in the nest, grow faster. The inference we drew from these facts was that these animals were doing their growing in the nest, that they were probably doing it fast—and that they were therefore probably not cold-blooded like most modern reptiles but warm-blooded like birds and mammals.

The evidence from this nest and these fragile bones said that dinosaurs seemed to have avian physiology, behavior, and growth patterns. It was a revolutionary finding.

The time was right, too. A few years before, John Ostrom and his then-student Robert Bakker had started exploring the idea that dinosaurs might be warm-blooded and highly active creatures. Exactly who first thought of the theory that dinosaurs were endothermic, or warm-blooded, has been hotly contested ever since. Both Ostrom and Bakker share the gift of seeing—and making others see—the past come to life. Ostrom is more often credited with the idea by paleontologists, perhaps because he has the advantage of being a beloved "grand old man" in the field who has always been associated with an impeccable academic institution, Yale University. In contrast, Bakker is younger, decidedly eccentric, and occasionally abrasive. Unlike Ostrom, Bakker has not settled into a prestigious faculty or museum position—his recent book, *Raptor Red,* gives the Tate Museum of Casper College as his academic affiliation—and he has far fewer publications in mainstream scientific journals than Ostrom. But Bakker's flair for conjuring up the past in words and drawings has led to considerable exposure of his ideas in the popular media. Priority in ideas, particularly between a professor and a student in the same institution, is a tangled web, impossible to unravel without breaking strands. Ostrom's and Bakker's disagreement led to bitter words and hard feelings that have not yet lost their edge. Perhaps the fairest reading of the situation was that endothermy in dinosaurs was an idea whose time had come, discussed by many and for which evidence was sought by several in the 1970s.

The evidence from Horner and Makela's dig went further than mere endothermy and seemed irrefutable. Their work showed that dinosaurs engaged in parental care of youngsters still in the nest. As Horner puts it,

> This was the first time that anyone had found a nest not of eggs but of baby dinosaurs, and the evidence seemed to me incontrovertible that these babies had to have stayed in the nest while they were growing and that one or more parents had to care for them. This kind of behavior, unheard of in dinosaurs, was probably the most startling discovery to come out of that dig. . . . it was in such severe contrast to the image of how dinosaurs were supposed to behave—laying eggs and leaving them, like turtles or lizards or most reptiles. If dinosaurs . . . had acted like birds and reared their young in nests, caring for them and bringing them food, this was a bit of information that would profoundly change our sense of what sort of creatures these ancient reptiles were.

In 1994, almost twenty years later, an international team from the American Museum of Natural History, George Washington University, and the Mongolian Academy of Sciences discovered the first theropod embryos ever found in the form of a nest full of *Oviraptor* eggs close to the time of hatching, at a 75-million-year-old Mongolian site called Ukhaa Tolgod. The eggs are carefully arranged within the nest, presumably by the mother, as birds do. The next year, in 1995, the same team was lucky enough to find the definitive specimen of a nesting dinosaur: an 80-million-year-old *Oviraptor*, seated atop a clutch of fifteen to twenty-two eggs. The *Oviraptor*, a small meat-eating dinosaur, was found in a very birdlike posture, with its legs tucked up tightly underneath its body and its arms turned back, encircling the nest in a protective manner. Horner may not have publicly voiced the thought, but he was entitled to a delicious moment of "I told you so."

Horner's work also implied that dinosaurs were *social:* a most unreptilian and entirely avian trait. In subsequent years, Horner and colleagues found better and better evidence of sociality. First there were *colonies* of dinosaur nests—complete with eggs and babies—with nests spaced one mother dinosaur's length apart from the next closest nest. It was just like modern colonies of puffins or flamingos, only with much larger babies, much, much larger mothers, and, probably, much louder squabbling. Second, Horner and colleagues soon turned up evidence of herds of dinosaurs aggregated by age group. Although there were a few diehard skeptics, the world as a whole simply stood back and applauded. By 1982, Horner was back home in Montana as curator of vertebrate paleontology at the Museum of the Rockies at Montana State University. In 1986, the University of Montana decided that their troublesome and erstwhile student had accomplished something, and awarded Horner an honorary doctoral degree. Also in 1986, the John D. and Catherine T. MacArthur Foundation followed suit, making Horner a fellow and recipient of one of their so-called genius awards. Much as the acclaim and awards have been welcome—and have made funding expeditions considerably easier—from Horner's point of view the real gain has been that all the hoopla has freed him to glean new insights into dinosaurs.

The evidence Jack Horner and his colleagues have mustered about complex social behavior in dinosaurs is eye-opening and just plain *fun.* Dinosaurs have been resurrected from huge, stupid, gray reptilian dullness to the brightly colored, honking, social, and swift-moving beasts we see them as today. Birdlike dinosaurs are endlessly fascinating and well-nigh irresistible. Thus recognizing the close relationship between birds and dinosaurs has illuminated dinosaur studies wonderfully. However, it has done nothing for birds and very little for *Archaeopteryx*. The crux of the problem comes back to one of perspective. In the words of Larry Martin, the Mesozoic bird specialist:

> The very first thing you do, when you look at a map or anything unknown, is that you recognize all the familiar landmarks that you've already seen. Now if you've never worked on birds, and you've worked on dinosaurs, when you look at *Archaeopteryx,* you're going to see the dinosaur. And if your primary work is on birds, then when you look at *Archaeopteryx,* you're apt to see the bird.

Although many paleontologists and ornithologists have studied the skeletal morphology of *Archaeopteryx,* Martin believes he is the only one who has also worked extensively on other fossil birds. It is a shocking claim. The fact is that most ornithologists are so closely focused on modern birds and their habits and adaptations that they have little time for the First Bird. It is in a sense the flip side to Jack Horner's strongly avowed disinterest in dinosaur extinction: "I don't give a damn how they died," he says. "I want to know how they *lived*." For their part, ornithologists don't care so much where birds came from as much as where they are *now*.

And where they are is fairly clear. The most obvious, exclusively avian trait today is that all birds have feathers. While there are other detailed anatomical traits unique to birds, many of these require considerable study and expertise to identify. Only a bird in the hand will tell you that its hollow bones are filled with air sacs, for example, or that it had a keeled sternum or a triossial foramen to perform the wing flip. Intuitive classifications are necessarily crude yet provide immense emotional satisfaction. But to trace the intricacies of evolutionary history, we need more: we need the scientific precision of anatomical detail. Only once these are known can we discover the appropriate taxonomic categories for and phylogenetic relationships among the organisms we observe. But should an animal be defined by what it *does*—its behavior and adaptations—or what it *is?*

Both dynamic actions or behavior and static taxonomy apply to the problem of defining a bird. What makes birds *birds*—what has shaped the anatomy and physiology of the Aves from head to toe, from skeleton to feathers—is dynamic: it is what they collectively do. Birds walk on their hindlimbs and fly with their forelimbs. They have invaded (evolutionarily) a unique adaptive niche, or way of life. No other group has ever accomplished this in quite the same way. In this sense, wings and feathers are among the essential attributes of birds, along with the innumerable specializations of breathing apparatus, thermoregulatory system, skeleton, muscles, nerves, and skin that accompany them.

A simple list of diagnostic features possessed by birds includes: 1) a bodily covering of feathers; 2) a toothless beak; 3) shoulders and elongated forelimbs incapable of providing terrestrial support and modified for flight (i.e., with a keeled sternum, a fused furcula, a triossial canal permitting a wing flip, and only three digits on the hand); 4) hollow bones filled with air sacs; 5) hindlimbs adapted for an obligate bipedal posture and gait, via changes in the pelvis and a reduction of the leg and ankle bones to form a tarsometatarsus; 6) remiges attached to the tail vertebrae, which are reduced in number to form a pygostyle; 7) a metabolism that is both endothermic (warm-blooded) and hypermetabolic (extremely active). From this perspective, the only answer to the question "What is a bird?" is "A consummate flying animal," but accuracy forces me to add, less satisfactorily, "of a particular type."

The second part of the answer is crucial. Lots of animals fly—butterflies, eagles, bats, and pterodactyls, to name only a few disparate types—but only some of them are birds. That is because the way in which a bird is adapted to flight is predicated upon its ancestry. Only certain pathways and adaptations are possible, given the starting point of the ancestral bird; a certain number of anatomical and physiological features are preselected by history, whose dictates are absolute. Thus, instead of evolution's being a broad river of endless possibilities, it is a series of branching rivulets of ever-diminishing size that rarely (if ever) recombine or reverse direction. Each turn, once made, eliminates many other possibilities, until the route becomes narrow and canalized. At this point, it is easy to cast an entire pathway as a series of deliberate choices reflecting intent, which it is not. The error of teleology is to mistake the present condition for a consciously chosen destination. The present is not a consciously chosen goal, nor is the evolutionary history of any group a trajectory except when viewed in retrospect. The answer "A bird is a consummate flying organism of a particular type" is unsatisfyingly vague unless it is informed by a more specific sense of what that "particular type" might be.

What type of creatures are birds? It is far easier to answer this question when it is applied to a single species than to a taxon as large as Aves. A species is a group of actually or potentially interbreeding populations that is genetically and reproductively distinct from other such populations. Thus ostriches in the San Diego Zoo may never see ostriches on the Serengeti Plains of Tanzania, but they are nonetheless members of the same species because they could interbreed if they met—and if those meeting were of opposite sexes and suitable inclination. Species are then grouped into larger categories known as genera, genera are clustered into families, families are united into orders, and orders are combined into classes. There are even finer divisions—subfamilies, suborders, and so on, to accommodate more complex gradations of relationship. The hierarchical and nested structure of these taxonomic categories, or taxa, can be best understood by taking a relatively uncontroversial example. Starting from the lowest or most exclusive category, all living humans belong to the species *Homo sapiens.* This species is one of several grouped under the genus *Homo,* which is in turn one of several genera grouped into the family Hominidae (humans and their ancestors since the split from the ape lineage). Grouped with the other taxonomic families (such as those representing the apes, the monkeys, and the lower primates such as lemurs, lorises, tarsiers, and bushbabies), humans are also members of the order Primates which, along with other orders of animals, constitutes the class Mammalia. As these taxonomic categories increase in scope, from species to class or from specific to general, each new step encompasses all previous steps; they are nested categories. Oddly enough, the differences among the major categories are so enormous that it is tricky to deduce which are most closely related to which others.

Giving an equivalent taxonomic placement for birds in either diagrammatic or written form is much more difficult because there is so much disagreement. Like primates, any particular living bird (such as the great horned owl) has a species indicated by a two-part Latin name *(Bubo virginianus),* a genus (the first part of the name, or *Bubo*), a family (Strigidae), an order (Strigiformes), and a class (traditionally called Aves). At each level, there are many more bird taxa than in the primate example, because birds are so much more abundant and diverse. There are also many more nested categories because the evolutionary picture of birds is so complex: species, genus, family, order, and class just don't adequately describe the reality. Also, with the rise of Hennigian cladistics, the practice of naming specific higher taxonomic categories (such as families, classes, and so on) has been largely abandoned in favor of the more neutral term "clade." Both a species and a class are examples of a clade. This practice makes determining the taxonomic equivalence of two clades almost impossible. It also symbolizes

the fact that cladograms—the branching diagrams of organisms' relationships that result from a cladistic analysis of the features of those organisms—are not phylogenies (diagrams of evolutionary pathways). Still, there
is generally good agreement about the classification of most birds up to the
point where they all fall into the clade Aves. The next largest category into
which Aves fits is controversial.

Many paleontologists exclude *Archaeopteryx* and some of the earliest birdlike species from Aves because, for example, they still have teeth and many tail
vertebrae, which have not yet been reduced into a birdlike pygostyle. In that
case, Aves plus these extinct forms can be grouped into a new and larger clade,
the Avialae. Avialae, in turn, is nested within a still broader clade, the Archosauria, which includes most of the animals that are colloquially called reptiles (excepting lizards, snakes, turtles, and a few extinct forms). Archosaurs
are recognized by a diagnostic suite of anatomical characters, eight of which
can be identified from skeletal remains (and thus on a fossil). Most of these
have to do with the structure of the skull, such as the possession of antorbital
fenestrae, which are specific holes through the skull below each orbit. A few
of the diagnostic features of archosaurs pertain to the rest of the body, such as
the possession of a fourth trochanter (a bony bump for the attachment of ma-

jor leg muscles) on the femur, or
thigh bone. Another of the diagnostic features of the archosaurs is
that their teeth are laterally compressed (long and narrow) and ser-
rated like a steak knife. Birds barely
squeeze into this category, by
virtue of having antorbital fenestrae, even though all living birds
lack teeth. Even *Archaeopteryx*'s
teeth are conical and unserrated.

The most inflammatory issue of
all in avian taxonomy is where,
within the archosaurs, birds originated. There are three possibilities,
because two dominant evolutionary trends have split the archosaurs
into three subgroups that might be
diagrammed as a sort of Y with

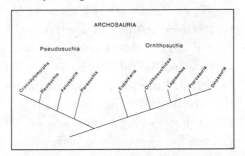

Figure 25. Jacques Gauthier's cladogram shows
the basic split within the Archosauria, the group
that includes crocodiles, birds, and their fossil ancestors. The left-hand branch, the Pseudosuchia,
evolves a more and more crocodile-like condition,
whereas the right-hand branch, the Ornithosuchia, evolves a more and more dinosaur-like
condition. The stem of the Y in this diagram represents the poorly defined group known as the
basal archosaurs or primitive pseudosuchians.

(Reprinted with permission; after J. Gauthier, "A cladistic
analysis of the higher systematic categories of the Diapsida," Ph.D. thesis, University of California, Berkeley,
1984.)

branches radiating from each arm of the Y (Figure 25). These, then, are the three possible nesting sites for the ancestry of *Archaeopteryx* and other birds: primitive pseudosuchians (the stem of the Y), the crocodylomorphs (one arm), and the dinosaurs (the other arm). The stem of the Y includes only a few known "basal" archosaurs that seem to be very primitive; they must have arisen either before the two main trends got going or so soon after the initiation of those trends that their alignment with either group is unclear. These orphans don't make a very convincing or cohesive group in and of themselves and comprise what might be called the "odds and sods" or wastebasket taxon of the Archosauria. For convenience, this assorted group has been called the Thecodontia or, for a somewhat smaller ragbag of creatures, the Pseudosuchia. The difficulty is that these names imply that either thecodonts or pseudosuchians are a valid and respectable taxon, which no one in paleontology really believes to be true. What the thecodonts or pseudosuchians are is "none of the above."

Jacques Gauthier, now of Yale University, was one of the pioneers in applying rigorous cladistic methodology to the question of bird origins, bringing new techniques of analysis to an old, old question. Not surprisingly, a number of his publications are coauthored with his former Ph.D. supervisor, Kevin Padian of the University of California, Berkeley. Padian, for his part, trained at Yale and was thoroughly familiar with Ostrom's ideas. Thus Padian and Gauthier loosely represent a metaphorical son and grandson of Ostrom and the relationship is apparent in their research. A major difference is that the younger scholars embrace cladistic methodology in a way that Ostrom never has. For his Ph.D. dissertation, Padian investigated different modes of flight, especially in pterosaurs, a topic quite similar to what Ostrom had intended to research when he was diverted by stumbling across the Teyler Museum *Archaeopteryx*.

Gauthier and Padian object strongly to vague terms like pseudosuchian on the grounds that the group has no biological reality. "In current usage," they wrote in a 1985 paper, "'Pseudosuchia' is a refuge for archosaurs of uncertain phylogenetic relationships that do not belong to any well-defined . . . taxon." If Gauthier and Padian sound a little like the members of an aristocratic club that a ne'er-do-well of unknown ancestry is trying to join, their point is nonetheless absolutely valid. Until substantially more fossil evidence is found, it will be impossible to resolve where these oddities belong in the evolutionary scheme of things. Neither the thecodonts nor the pseudosuchians are a valid taxonomic group that is monophyletic, or descended from a single ancestor. As a consequence, they violate the basic assumptions underlying Hennigian cladistics. Since pseudosuchians and thecodonts cannot be assessed by one of the primary tools of modern paleontology, Gauthier and Padian feel that the only reasonable recourse is to omit them from any rigorous analyses. It is an awkward situation, effectively denying the very existence of perfectly good specimens because the background knowledge of the groups they represent is so poor. The alternative is even more distasteful: applying cladistic methods inappropriately, which will guarantee a meaningless result. For this pair, and most other paleontologists who have endorsed Hennigian cladistics, the pseudosuchian hypothesis is flatly intractable. Nonetheless, the pseudosuchians and the thecodonts are still sometimes touted as possible bird ancestors. The strength of this argument is that because they are so primitive it is easy to envision them evolving in any number of directions. They may not be a secure roost for the origin of birds, but at least they cannot be eliminated from candidacy.

Fortunately, the arms of the archosaur Y are more specialized and less problematic taxonomically. One is the lineage that leads toward Crocodylomorpha, or the crocodile-shaped animals; the other leads via pterosaurs to dinosaurs and—according to many but not all paleontologists—to birds. These two arms represent opposite evolutionary trends that, in hindsight, can be caricatured as "toward crocodiles" or "toward dinosaurs" that developed out of the primitive, basal group of archosaurs. These trends are particularly clear in the way the ankle joints are built in the different subdivisions of the Archosauria. Basal archosaurs have a simple, primitive, hingelike ankle, in which the axis of bending occurs between the ankle bones (the main ones are the astragalus and the calcaneum) and the foot itself. The Crocodylotarsi, or crocodile ankles, is the name given to the lineage that gets more and more crocodile-like over time, ending in the Crocodylomorpha. This lineage abandoned the primitive ankle arrangement in favor of an arrangement in which bending occurs *within* the ankle itself—between the astragalus and the calcaneum—and not between the ankle and the foot. In a crocodilian-type ankle, a bony peg on the astragalus fits into a socket on the calcaneum and keeps the two in proper alignment as they rotate during bending.

The other lineage, the Ornithosuchia, gets increasingly dinosaur-like or birdlike and develops two sorts of advanced ankle joints. The more primitive members of the Ornithosuchia have an ankle joint called the crocodile reversed, in which the plane of bending falls within the ankle, as in crocodiles, except that the calcaneum has a bony peg that fits into the astragalus rather than the reverse. The more advanced Ornithosuchia, a group sometimes called the Ornithodira, have an entirely different arrangement. There is no plane of bending between the astragalus and calcaneum of the Ornithodira. Instead, these two ankle bones are virtually fused to each other and to the tibia (the large lower leg bone), making a single, inflexible leg-and-ankle unit. The astragalus-and-calcaneum unit is bound to the tibia by a special bony flange called the ascending process of the astragalus. This tongue of bone arises from the front of the astragalus and rises upward to fit into a special notch on the front of the tibia, where strong ligaments hold it in place. This arrangement is found in dinosaurs (the Dinosauria), birds (Aves), birds plus the early birds (Avialae), pterosaurs (Pterosauria), and a few other extinct forms (Figure 26). Perhaps ankle structure doesn't seem a likely candidate for the most important way to recognize birds and their close relatives, but a hindlimb dedicated to bipedal locomotion is a crucial aspect of their adaptation to flying. Clearly at some point bird ancestors must have been bipedal, or else the forelimb could not have been freed to develop as a wing, and the evolution of upright posture necessitated major rearrangements of the foot, ankle, knee, and hip to ensure stability and weight-bearing. Becoming bipedal left lasting changes in all of the bones of the hindlimb in dinosaurs and in birds.

Why then aren't birds simply a subdivision of dinosaurs? Many scientists

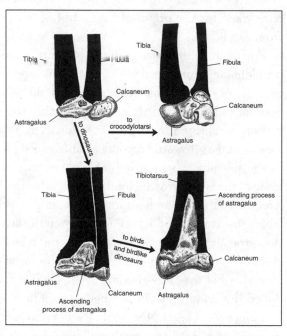

Figure 26. A major distinction between the various lineages of archosaurs lies in the structure of the ankle joint. The primitive condition is shown in the upper left. In crocodylomorphs (upper right), bending occurs between the astragalus and the large calcaneum, within the ankle itself. In the Ornithosuchia, the calcaneum and astragalus become more tightly connected to each other and to the tibia; bending occurs below the inflexible unit formed by the tibia, astragalus, and calcaneum.

believe they are. They see living dinosaurs perching on our back-yard bird feeders and gracing our Thanksgiving tables. As Gauthier says, "Birds are as much dinosaurs as humans are mammals." This is a strong and appealing hypothesis, particularly because it gives us some powerful perspectives from which to understand dinosaur biology, as Jack Horner's work has shown. But a serious rival hypothesis focusing on the Crocodylotarsi must also be evaluated. The Crocodylotarsi are united by

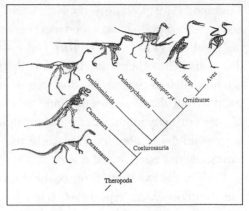

Figure 27. Within the Dinosauria, the groups most closely related to birds lie among the theropod dinosaurs. This simplified cladogram shows the relationships among some of the major groups of dinosaurs and birds.

their common ankle structure, although the members of this clade are diverse in their locomotor adaptations. For example, the aetosaurs were armored, blunt-snouted quadrupeds, while the rauisuchians were dominantly quadrupedal but could manage a specialized erect gait. The popsaurids were another group of Crocodylotarsi that were bipedal carnivores, while the species in the Crocodylomorpha (the crocodile-shaped animals) themselves were small, bipedal, and terrestrial early in their evolution; only later in time did species in the Crocodylomorpha develop into the large, low-slung, water-adapted beasts with which we are now familiar. Small, bipedal crocodiles– difficult as they may be to imagine—sound suspiciously like reasonable candidates for avian ancestry, too. And those that favor this hypothesis have found many details to criticize in the theropod dinosaur hypothesis.

The history of theories about avian ancestry shows that first one and then another has been in vogue. Changes in the general consensus have been effected by new fossil finds. The initial discovery of *Archaeopteryx* moved O. C. Marsh, a Yale paleontologist, to suggest dinosaurs as the ancestor of birds; nearly simultaneously, Thomas Henry Huxley in London suggested and elaborated the same hypothesis in several papers that compared *Archaeopteryx* with *Compsognathus*, a theropod dinosaur of similar size and preservation to *Archaeopteryx*, and both with ostriches. But theropods are only one of three divisions within the Dinosauria that are recognized today (Figure 27). The

Theropoda are predatory dinosaurs and include velociraptors and tyrannosaurs; theropods are grouped together with the Sauropodomorpha, including plant-eaters like *Apatosaurus,* to form the Saurischia, or "lizard-hipped" dinosaurs. The other main division is the Ornithischia, or "bird-hipped" dinosaurs, such as the plant-eating stegosaurs, hadrosaurs, or ceratopsians. In pointing out the close resemblance between birds and theropod dinosaurs, Huxley's primary intent was to connect dinosaurs to modern forms. Although he may not have intended to place birds accurately in the context of extinct forms, the net effect was the same.

Early in the twentieth century, the dinosaur hypothesis was superseded by the pseudosuchian hypothesis, again because of new fossil finds. Robert Broom, an irrepressible Scottish paleontologist, had been excavating and describing bipedal, primitive archosaurs (pseudosuchians) and mammal-like reptiles from the Karroo region of South Africa. With the unerring instinct for big, catchy ideas that characterized his entire career, Broom proposed the pseudosuchians as the group ancestral to birds in 1906 and again in 1913. It was a thoroughly satisfactory way to make sure everyone paid attention to his new finds. Too, his pseudosuchians were so primitive that they were reasonable candidates for the ancestry of all birds. Broom's ideas and, in particular, his description of *Euparkeria,* a pseudosuchian dinosaur, had a tremendous influence on the debate over avian origins and shifted opinion away from the dinosaurs as bird ancestors. The pseudosuchian hypothesis maintained its ascendancy for many years in large measure because of a very influential book published in 1926, Gerhard Heilmann's *The Origin of Birds.* This volume stood for many years as the definitive work on the question of avian ancestry. In it, Heilmann assessed various groups in turn as possible "nests" from which birds might have hatched, including pterosaurs. These he found to resemble birds only superficially, pointing out that, as pterosaurs had no clavicle (the unfused version of the furcula), they could not possibly be ancestral to birds, which possess furcula. As Larry Witmer, a modern paleontologist, has observed, this is a very revealing statement. He writes:

> This quotation illustrates a guiding principle of Heilmann's book—i.e., Dollo's law of the irreversibility of evolution: that which is lost cannot be regained. Before beginning his analysis of the fossil groups, Heilmann asserted (p. 140) that "when strictly adhering to this law, we shall find only a single reptile-group can lay claim to being the bird-ancestor."

Despite many detailed resemblances between theropod dinosaurs and birds observed by Heilmann, he was forced to rule the former out of avian ancestry because he believed they lacked clavicles (a point contradicted by more recent discoveries). Pseudosuchians win out as Heilmann's favored candidate for bird ancestry because "nothing in their structure mitigates against the view that one of them might have been the ancestor of birds." In other words, Heilmann favors pseudosuchians as bird ancestors simply because there is no negative evidence—because nothing eliminates them from further consideration—not because of strong, positive resemblances. Saying that birds may have arisen from the pseudosuchians is a little like saying the first bird hatched out of the egg of something unknown; it is probably true, but it is not a very helpful observation.

Another interesting "discovery" began to wield serious influence over paleontologists' views of avian ancestry at about the same time that Broom's new fossils were thrusting pseudosuchians forward. In 1907, William Pycraft of the British Museum (Natural History) created a purely hypothetical bird ancestor, known as "Pro-aves" (later spelled "Proavis"). Pycraft constructed a scenario for the origin of avian flight that featured this original, quadrupedal, lizard-like fossil (of which he had no evidence) that leaped and parachuted from tree to tree as a prelude to flight. The name and concept of "Pro-aves" caught on, but its only constant attribute has been its structural position as ancestor to all later birds. Soon after Pycraft's imagination hatched "Pro-aves," Franz Nopsca, an eccentric Austrian baron, drew his own version of the mythical ancestor that he called "Pro-Avis" (Figure 28). It is a rather charming, running creature with scales on the trailing edge of its arms that, according to Nopsca, would later evolve into true feathers. Resurrecting an old idea about the origins of flight, Nopsca wrote confidently in italics:

Figure 28. The hypothetical ancestor of birds, "Proavis," is shown as imagined in 1907 by the eccentric Austrian baron Franz Nopsca. No fossils of "Proavis" have yet been found.

> *Birds originated from bipedal long-tailed cursorial reptiles which during running oared along in the air by flapping their free anterior extremities. . . .* By gradually increasing in size, the

111

enlarged but perhaps still horny hypothetical scales of the an-
tibrachial [forearm] margin would in time enable the yet carniv-
orous and cursorial ancestor of Birds to take long strides or
leaps, much in the manner of a domesticated Goose or a Stork
when starting, and ultimately develop to actual feathers.

Reconstructions of fossil-free ancestral birds became popular. In 1915,
William Beebe described another hypothetical ancestor, "Tetrapteryx,"
which had feathers on both fore- and hindlimbs. In turn, German scientists
started to write of the "Urvogel," the primeval or essential bird. Of them all,
only "Proavis" took on a life and personality of its own. "Proavis" is still universally recognized as *the* ancestral bird—the one that preceded *Archaeopteryx,* and the one that truly gave rise to all later birds. Although its
iconic status is not quite as hallowed as that accorded to *Archaeopteryx,*
"Proavis" is still dear to a great many paleontologists and ornithologists.

This fact was to have peculiar consequences for a modern paleontologist,
Sankar Chatterjee of Texas Tech University, who had the audacity to endow a
possible fossil bird with a closely similar name, *Protoavis.* Based on a highly
fragmented partial skeleton, Chatterjee announced that he had found the
earliest bird: one that was not only earlier than *Archaeopteryx* (by about 75
million years) but also more similar to modern birds than *Archaeopteryx.*
Following upon a press release from Chatterjee's sponsors, the National
Geographic Society, the claims for a new First Bird received extensive media
attention, a point that drew sharp criticism from some fellow paleontologists,
who wished Chatterjee had waited until his scientific publications were prepared and reviewed by his peers. His was a very sensational claim to be
spread across the newspapers before other scientists had a chance to evaluate
Protoavis. "I never went to the press," says Chatterjee softly, in his own defense. "I did not do the press release. I never called any journalists." It was
his sponsors who chose to announce his findings when they did and Chatterjee had no reason to refuse. Aside from its unfortunate early publicity and the
bad feelings engendered by too much media hyperbole, *Protoavis* has remained problematic. It is at best an extremely difficult specimen to interpret.
After seeing the specimen, Gauthier described it informally as "smushed and
mashed and broken"; another paleontologist, Tim Rowe of the University of
Texas, referred to it as "real roadkill." Even Larry Martin, who is less skeptical of Chatterjee's claims than some, admits that after a week of examining
the specimen in detail, he is still uncertain that Chatterjee has correctly iden-

tified all of the fragments in terms of the bone of origin. Larry Witmer, who has also examined the specimen at length, says simply that the specimen is "very scrappy." He adds, "I have to admit that I'm doubtful of its avian status, but I'm willing to admit that it does have some birdlike features. The question is, how many birdlike features does it have, how birdlike are they, and what do they mean?" The best course, suggested by Luis Chiappe of the American Museum of Natural History in a recent review, is that *Protoavis* should not be considered relevant to avian evolution until better specimens are available. At present, *Protoavis* must be considered possibly real and possibly just another variant of the hypothetical "Proavis."

The pseudosuchian hypothesis (enlivened by "Proavis" and other flights of fancy) ruled until the early 1970s. Then, in 1972, Alick Walker of the University of Newcastle-upon-Tyne first suggested that birds and crocodylomorphs shared a common ancestor. He based his theory largely on the discovery of *Sphenosuchus,* a Triassic form he saw as intermediate between crocodilians and birds. Unfortunately, Walker's theory is easily mischaracterized. He did not suggest that birds are descended from *crocodiles* —seemingly a patently absurd proposition—but rather from *crocodylomorphs.* The problem is actually one of the failure of colloquial vocabulary, for there is no good, broadly understood term with which to designate "a group of largely extinct animals similar in overall body shape to crocodiles or alligators." Walker's position was strengthened greatly by the support of Larry Martin and some of his students. They found resemblances among avian and crocodilian quadrates (a bone of the jaw that articulates with the skull) and turned up interesting, detailed evidence that crocodylomorphs have a fundamentally different sequence of tooth implantation and replacement than did dinosaurs. Judging from *Archaeopteryx,* birds group strongly with crocodylomorphs on the basis of their dental apparatus, too.

This dental evidence forms the keystone of Martin's beliefs about avian ancestry. Says Martin,

> My viewpoint on the origin of birds is that birds were one of the first clades to come off the Archosauria—that they may have come off before socketed teeth were established. . . . An origin that early would have to drive the origin of birds back to the early Triassic; it might even have been that they split off in the late Permian. Now this would be sort of interesting because that was the point where for the first time we have good evidence

113

that we have arboreal animals. That's when we have all of these gliding animals that were sort of lizard-like, but not maybe real lizards. . . . That was the time when a lot of flying animals and arboreal animals were evolving, and I suspect birds rolled off at about that time. They may have had a close common ancestor with whatever gave rise to crocodilians at that time. But I have never argued nor do I argue now that birds are derived from crocodiles.

Just prior to Walker's revival of the crocodylomorph hypothesis, John Ostrom had made two fateful discoveries. In 1969, in Montana, he found a new type of theropod dinosaur that he named *Deinonychus antirrhopus,* with a long, low skull, a formidable array of teeth, and slashing, sickle-shaped claws on its hands and feet. It was a small, agile, and fiercely predatory creature with grasping hands that helped convince Ostrom that dinosaurs must have been warm-blooded. Some paleontologists now consider *Deinonychus* a type of *Velociraptor,* the terrifyingly swift and cunning predator in the film *Jurassic Park,* although others consider them separate but closely related. Studying the anatomy of *Deinonychus* prepared Ostrom psychologically for recognizing its similarities to *Archaeopteryx,* which he found hidden among the pterosaurs in the Teyler Museum in Holland in 1970.

As a consequence of this two-part discovery, Ostrom began to revive Huxley's dinosaur hypothesis of bird origins. Birds, he argued with the passion of a sudden convert, are so like small, theropod dinosaurs that an unfeathered early bird specimen could easily be mistaken for such a dinosaur. *Deinonychus* served as a compelling ancestral prototype for "Proavis" in his mind. Like others before him, Ostrom was forced to resort to conjuring up an as-yet unknown species as the specific ancestor because even *Deinonychus* was not perfectly equipped as an avian ancestor. But what he did see was a small, bipedal animal, with a long, stiff tail, a foot with three forward-facing toes and a reversed big toe for grasping, an enlarged hand with sharp, raptorial claws, and sharp teeth adapted for flesh-eating. In these regards, theropods are very like unfeathered *Archaeopteryxes.* Echoing Huxley's sentiments one hundred years earlier, Ostrom observed, "Were it not for the preserved impressions of feathers in *Archaeopteryx,* those specimens would have been classified as coelurosaurian dinosaurs. Also, had it not been for the feather impressions, *Archaeopteryx* just as certainly would have been pictured as a cursorial, bipedal predator, not an arboreal glider." Combined with the engaging idea that dinosaurs may have been warm-blooded, Ostrom's hypothesis rapidly gained widespread support among professional and amateur dinosaur lovers.

With the diagnostic features of birds specified, and the possible candidates (dinosaurs, pseudosuchians, crocodylomorphs) enumerated, tracing the ancestor of birds should be a relatively straightforward task. It should be a matter of comparing suites of features until a good match is found, with *Archaeopteryx* as a tangible clue to the sequence in which the different adaptations developed (Figure 29). But the task isn't quite so simple in reality. First, many of the scholars involved in the search for the origin of birds and bird flight flatly disagree about what the characters of various forms are. Martin in particular has challenged the theropod hypothesis on several grounds. One of the weaknesses of his alternative, crocodylomorph hypothesis is that neither Martin nor any of its other proponents have yet produced a rigorous cladistic analysis of the features they see as linking birds and crocodylomorphs. They have repeatedly criticized the cladistic features analyzed by dinosaurologists such as Gauthier and Padian, but have not countered with a comparable display of their own evidence. This renders their hypothesis more an "antidinosaur" theory than a "procrocodile" one.

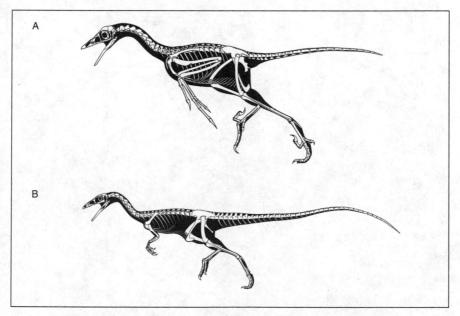

Figure 29. The skeletons of (A) *Archaeopteryx* and (B) *Compsognathus*, a small theropod dinosaur, show many resemblances. Historically, two of the seven specimens of *Archaeopteryx* were initially misclassified as *Compsognathus*, even though *Archaeopteryx* has proportionally longer arms and hands.

Another complicating factor is that *Archaeopteryx*—the focus of nearly all of the comparison and discussions—may have been the First Bird but it may not have been directly ancestral to all later birds. Features of its hindlimb and skull are so specialized that it is usually placed on an evolutionary side branch, albeit the earliest avian side branch known. Strictly speaking, this means that *Archaeopteryx* is not really a bird in taxonomic terms. Yet while John Ostrom and Larry Martin agree on almost nothing about the origin of birds and bird flight, they both cry "Oh, yes!" with conviction when asked if *Archaeopteryx* is a bird, colloquially. It is, they agree, a *wonderful* bird. And there the agreement ceases.

Chapter 5.

A Bird in the Hand

All the detailed wizardry that has been applied to understanding the flight of modern birds gives us a good basis for evaluating the features of ancient ones. *Archaeopteryx,* especially, is so well-represented by such superb specimens that its abilities, ancestry, and anatomy ought to be clear—and yet there is vibrant debate. I found myself asking, "Why?" With a bird (or parts of seven) in the hand, how can there be so little agreement about *Archaeopteryx?* At first, I suspected that the scientists were simply talking past each other, like small boys involved in an argument that has deteriorated to the "Is too!" "Is not!" stage. But as I investigated more, I found that there are good reasons for the persistence of disagreement. Yes, sometimes the tone of the debate degenerates into unsupported, personal convictions. But more often the morphology of *Archaeopteryx* is genuinely ambiguous because the ends of its bones were cartilaginous and didn't fossilize, and even the preserved bony parts may be twisted out of their true anatomical position or crushed and distorted by the process of fossilization and the heavy hand of Time. Sometimes, too, the curators' reluctance to fully prepare *Archaeopteryx* out of its matrix obscures anatomical details while preserving others, like the precious feather impressions. The specimens are, in this regard, almost *too* good to be maximally useful.

But the paleontologist in me cannot help but argue with this analysis of the problems of *Archaeopteryx*. After all, *Archaeopteryx* is known from seven partial skeletons; the sample is a veritable treasure trove by paleontological standards. Fossil vertebrates are rarely better known than this. Besides, would eight skeletons resolve the debates—or nine? Perhaps. The most convincing evidence that more material will resolve the debates comes from the most recent *Archaeopteryx* to be discovered, which has significantly changed views of the species' abilities and adaptations. The Solnhofer Aktien-Verein specimen has a bony sternum, which all of the other specimens lack. It is thus possible that this individual may represent a different species in the same genus, as Peter Wellnhofer has recognized in naming it *Archaeopteryx bavarica,* not *A. lithographica.* The specimen is also unusual in that it is a very small but apparently adult individual. Thus the Solnhofer Aktien-Verein *Archaeopteryx* is a case in point, showing that new specimens may provide new information, even about anatomical matters that were regarded as settled (like the "fact" that *Archaeopteryx* did not have a bony sternum). But the more pressing problem is not discovering the factual state of the anatomy but arriving at its appropriate interpretation.

Two main areas of the body—the wrist and hand, the foot and ankle—are the most contentious points and herein lies the most important evidence for, or against, the dinosaurian, crocodylomorphian, or thecodontian origin of birds. Embedded in the heart of each controversial point is a single principle fundamental to comparative anatomy—a topic so important that it overrides all other concerns in evolutionary biology. The subject is known as homology, and thousands of pages of effort have been expended trying to define and explain it.

The simplest definition is that homology means the quality of being alike or fundamentally similar because of commonness of descent. The idea of homology underlies the simple dictum "Like must be compared with like," and the proverbial objection to comparing apples and oranges. To find out anything useful about evolution and adaptation, a jaw must be compared to a jaw and a vertebra to another vertebra. In other words, homologous structures in two animals are the same entity, changed only by the dictates of their separate evolutionary courses. Even though they look superficially different, a human hand, a bird's wing, and a whale's flipper are all homologous in that each represents an evolutionary modification of the same ancestral limb structure. A formal and oft-cited statement of the principle of homology is that "attributes of two organisms are homologous when they are derived from an equivalent characteristic of the common ancestor" that links them. Recognizing the principle of homology is much easier than recognizing a particular homology, especially when two organisms that are only distantly related are being examined. Homologous structures in different organisms will develop from the same embryonic substances according to basically similar genetic instructions; for example, true teeth will always be made of dentine and enamel and, however much it may look like a tooth, a structure in the jaw made of something else is unlikely to be homologous to a tooth. This is why arguments over whether two structures are homologous or not often have recourse to embryological evidence, which may make the fundamental attributes of an unknown structure clearer. (It should be noted that "homology" is used in a rather different way by molecular biologists in comparing two genetic sequences. Their use of homology does not necessarily imply descent from a recent, common ancestor but speaks merely of sequence similarities.)

In anatomy and paleontology, the trick in recognizing homology is to distinguish it from another type of resemblance, known as analogy. (Both terms were coined by Richard Owen, Huxley's antagonist in the early tussles over *Archaeopteryx*.) Analogy refers to a similarity of appearance due to a similarity of function, not common descent. As Owen explained, anatomical parts are analogous if they perform similar functions (are functionally equivalent) but are derived from different origins. A commonly cited example of analogous parts is the wing of a bat, a bird, and a butterfly. All are airfoils essential to producing lift for flight, but the three are formed from very different structures and are built in very different ways. They are not the same structure evolutionarily, although they are the same functionally.

In practice, homology is either fairly simple to recognize by close inspection of the structures in question or cursedly obscure. Étienne Geoffroy Saint-Hilaire, a French anatomist who helped establish the basic procedures and ideas of comparative anatomy, proposed two means of recognizing homology in 1818. First, he argued for the power of the *principle of connections,* which states that an anatomical structure may be modified, reduced, enlarged, or eliminated by evolution, but it cannot be transposed. Thus, in vertebrates, the humerus will always lie between the scapula (shoulder blade) and the radius and ulna (the bones of the forearm), whatever shape the humerus may assume because of functional demands. A humerus, no matter how it has been modified by natural selection, will never lie at the tip of the finger, for example. A bone in that position must be a phalanx and cannot be a humerus. Second, he proposed that what he called "composition" was an essential clue. Homologous structures will always be composed of the same kind of elements—a bone will be made up of bone, for example, usually with a developmental precursor in cartilage, but it will not be made of muscle tissue or dental tissues. Once again, embryological evidence is often useful in clarifying whether ambiguous structures are analogies or homologies.

The debates over the origin of birds and bird flight are an object lesson in the dangers and difficulties of confusing homologous parts and analogous ones. This fundamental problem is the source of the debate over the differences and similarities in the wrists, hands, ankles, and feet of *Archaeopteryx* and its putative ancestors.

One of the cornerstones of Ostrom's hypothesis that birds evolved from theropod dinosaurs is the homology that he sees among the wrist bones of *Deinonychus,* those of birds, and those of *Archaeopteryx* (Figure 30). His critics, notably Larry Martin and Sam Tarsitano, argue that the same bones are present in modified form in the wrists of *Archaeopteryx* and modern birds, but that the bones in the wrist of *Deinonychus* and other theropods are different and have been misidentified by Ostrom. If this is true, then the dinosaurian wrist is not homologous to the avian one. Nonhomologous parts cannot be evidence of an ancestor-descendant relationship and may, in fact, refute that possibility.

To evaluate these conflicting views, I needed first to consider the general plan of the vertebrate wrist that is shared by all birds, mammals, and reptiles. The primitive or original situation is to have a wrist with as many as nine small carpal bones, aligned in two rough rows. In the evolution of various lineages, some of these bones are lost and others are fused to their neighbors, so that the end result is fewer than nine wrist bones. Saint-Hilaire's principle of connections would seem to be the appropriate aid to identifying which bones are which, except that the case is complicated by the extreme degree of evolutionary modification. *Deinonychus* and *Archaeopteryx* each have only two remaining wrist bones in a single row. That they are wrist bones is shown by the fact that they lie, as expected, between the radius and ulna of the forearm and the metacarpals of the hand. But since so many of the wrist bones have been lost, there is no obvious way to tell which they are, nor is it clear whether they started out in one or two separate rows. The issue could perhaps be resolved by an embryological study of both avian and dinosaur wrist development, but at present such a study is impossible to conduct. A number of embryos of the appropriate species *and* at the right stages of development and state of preservation would be needed; they are simply not available. Even if they were, the evolutionary change from nine wrist bones to two is so major that even detailed embryological studies might not clarify what has happened and where the homologies lie.

And yet the homologies or analogies of the wrist bones are important because of their specialized function. Birds can bend their wrists sideways in a gesture that in humans would bring the little finger up against the side of the forearm. Birds do this to fold their wings and tuck their feathers up close to the body, to keep them from harm or from interfering with other activities. This ability is part of the automatic folding and unfolding mechanism that lets birds furl or open their wings with minimal muscular effort, and is very important in flying. Ostrom and his colleagues have no difficulty imagining the evolution of such lateral bending, which they see as very similar to the movements of the theropod wrist in a grasping or scooping action that might have been performed with feathered hands. Rick Vasquez, a paleontologist trained at Yale, has studied the movements of the avian wrist more thoroughly than anyone

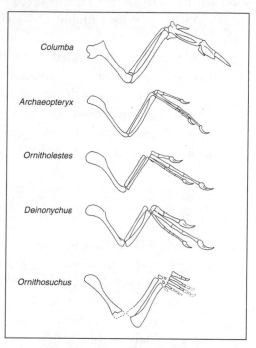

Figure 30. John Ostrom sees close resemblances in the structure of the forelimbs of birds (such as *Columba*, the pigeon), *Archaeopteryx*, and theropod dinosaurs like *Ornitholestes, Deinonychus,* and *Ornithosuchus.* These similarities persuade him that theropod dinosaurs are ancestral to birds.

else. Studying *Archaeopteryx*, Vasquez sees almost no signs of the specializations of the modern avian wrist that permit wing folding or the rotation of the wing that is essential to a successful recovery stroke during flapping. In his view, *Archaeopteryx* was simply "incapable of executing the same kinematics displayed by modern birds during flapping flight." The wrist just isn't built to let *Archaeopteryx*'s wings move and flap the way those of modern birds do. In particular, Vasquez is convinced that *Archaeopteryx* could

121

not possibly rotate its hand and fold its wing to tuck its feathers up against the body.

Alan Feduccia, an internationally known biologist at the University of North Carolina, challenges this conclusion. Feduccia trained as an ornithologist, earning his Ph.D. from the University of Michigan with a study of oven birds and tree creepers. He approaches *Archaeopteryx* strictly as an ornithologist, one who never intended to participate in fossil debates. "It was back in the 1970s that I got involved with *Archaeopteryx*," he explains, speaking in a quiet voice with just a trace of a southern accent.

> I had pretty much decided to write a book—a synthesis—on what we know about bird evolution, concentrating more on the modern groups, especially the perching birds. And I discovered a big void in our knowledge about the origin and evolution of birds and avian flight. The more I looked into it, the more I did not like what I was seeing: none of it made sense.

One of the problems was the idea, then growing rapidly in popularity, of hot-blooded dinosaurs as the ancestors of birds. Feduccia continues,

> I've studied bird skulls for thirty years and I don't see any similarity whatsoever to dinosaur skulls; I just don't see it. I'm of the opinion that the whole hot-blooded dinosaur thing is very seriously flawed. Part of the problem is, for ornithologists, *Archaeopteryx* is a bird, not a dinosaur. For those of us who study feathers, feathers are the most complex integumentary derivatives of vertebrates. They are completely aerodynamic in structure. Why evolve this on an earthbound dinosaur? It doesn't add up somehow. I think people are trying to turn dinosaurs into birds. This is more of a sociological phenomenon than a science.

And it doesn't make sense to him that *Archaeopteryx* has wings that wouldn't fold up against the body; it is virtually inconceivable. He points out that the Berlin specimen has flexed wings, proving that flexion is possible; that the fossils do not preserve the cartilaginous parts of the wrist, only the bones, and so part of the anatomy is missing; and that the feathers of *Archaeopteryx*, like those of any modern bird, are arranged in order to "fold up like an old-fashioned fan." The first two of these objections are not entirely convincing.

The Berlin specimen's wings are flexed, but only partially; the feathers are clearly not tucked up against the body. As for the wrist of *Archaeopteryx*, it is true that only the bony wrist can be evaluated. However, at least some fossil birds (like *Apatornis,* an Upper Cretaceous species from North America) show the adaptations for wing folding on their wrist bones, so *Archaeopteryx* could have done the same, too, if it had those adapta-

Figure 31. An early scenario by John Ostrom shows "Proavis" (top) as a ground-dwelling theropod using its wings as a net to capture insect prey. "Proavis" would then evolve ever larger wings until it arrived at the condition of *Archaeopteryx* (bottom). Ostrom later agreed with his critics that feathers would make a poor insect net and suggested that "Proavis" caught prey with its teeth, using its incipient wings to prolong leaps into the air.

tions. Still, it is theoretically possible that cartilage changed the topography of *Archaeopteryx*'s wrist substantially. And although *Archaeopteryx*'s feathers do seem designed to fold up like a fan, this elegant metaphor is constructed with the assistance of hindsight, looking back on the evolution of flight from the present perspective. While wing feathers fold up like fans in modern forms, they may not always have done so.

John Ostrom doesn't welcome Vasquez's conclusions any more than Feduccia does, although Ostrom bows to Vasquez's authority in this matter. Perhaps optimistically, Ostrom sees the first hints of some of these anatomical adaptations in primitive form in *Archaeopteryx* and in theropods like *Deinonychus. Deinonychus* is a member of a group sometimes called maniraptors, or hand predators, who are believed to have used their long, clawed hands in a forward and downward scooping motion to seize prey and bring it toward their mouths. Like these dinosaurs, *Archaeopteryx* had sharp, pointed claws on its fingers that might have been useful in impaling or grabbing prey (Figure 31). In 1979, Ostrom constructed an evolutionary scenario, taking a small theropod as the ancestor of *Archaeopteryx*:

> If these animals used the three-fingered hand to seize ever smaller and smaller prey, any modifications that would improve insect-catching ability would be highly advantageous. Is it possi-

ble that the initial (pre-*Archaeopteryx*) enlargement of feathers on those narrow hands might have been to increase the hand surface area, thereby making it more effective in catching insects? Continued selection for larger feather size could have converted the entire forelimb into a larger, lightweight "insect net.". . . The "proto-wing" at first may have been used in sweeping movements to flush insects concealed in ground cover or low vegetation. . . . From simple sweeping arm movements, to flapping swipes at leaping or flying insects, to flapping leaps up after escaping insects seems to be a logical sequence within the insectivorous precursors of *Archaeopteryx*.

But *Archaeopteryx* has feathered hands, which would make these sweeps and swipes rather awkward. A few years later, Larry Martin voiced a widespread feeling that Ostrom's scenario was physically improbable:

> The suggestion that the feathers were an "insect net" is also hard to support. The reason that an insect net is a *net* is to permit air to pass through it. Without this special quality the insects would be blown out of the structure by the rush of air. A similar problem would have arisen for the hapless proavis as it sought to trap its dinner, unless the structure of the feathers provided a free passage of air through their fabric.

Air cannot pass freely through a feather, which is why it is useful in flight but could not have been useful as an insect net.

Also, according to careful reconstructions by both Burkhard Stephan of the Museum für Naturkunde der Humboldt-Universität and Siegfried Rietschel of the Staatliches Museum für Naturkunde in Karlsruhe, the feathers on the wings of *Archaeopteryx* are long and would have extended beyond the tips of all of the fingers (Figure 32). The innermost finger was much shorter than the others and may have been freed of involvement in the feathered wing. The middle-sized middle finger and the long, outermost finger both were feathered, not naked. While the claws were not covered in feathers, most or all of the fingers themselves were, making them much broader structures than simple fingers. When the wing or arm was stretched out, the clawed fingertips were not the most distal part of the hand: the tips of the *feathers* were. As I looked closely at these reconstructions, I could see immediately that the feathered finger probably could not be used to grasp prey, for the feathers would be bent or broken, especially if the prey struggled. The anatomy is all wrong for *Archaeopteryx* to have functioned as either a maniraptor or an insect netter, although *Archaeopteryx* may well have been predatory and cursorial.

Another objection to the *Archaeopteryx*-as-maniraptor that Ostrom espoused in his earlier papers was raised by a trio from Northern Arizona University consisting of Gerald Caple, a chemist, Russell Balda, an ornithologist, and William Willis, a physicist. When they modeled the effects of lunging after insect prey with feathered hands or proto-wings, as Ostrom had theorized in his scenario, they found that the action would have upset *Archaeopteryx*'s aerodynamic equilibrium and pitched it forward onto its beak. Clearly this is not a viable way to chase insects. However, the team found that if *Archaeopteryx* or its ancestors caught prey between their sharp little teeth, even incipient wings would have helped to control the direction and duration of predatory leaps. But using small or incipient wings to prolong leaps is an improvement only if the main area of the wing is placed far enough from the center of gravity of the body; if the broadening of the arm into a wing begins at the armpit, there is no improvement in stability or steering. If this model is correct, then, you would expect the hand to enlarge to form a wing before the upper arm and forearm lengthen and broaden. This is indeed the case in

Figure 32. After careful study of the individual feathers of the Berlin *Archaeopteryx*, Siegfried Rietschel reconstructed its outstretched wing. Because the fingers were feathered, there is no position in which the claws were the outermost structure.

Archaeopteryx, which suggests that the model is reasonably accurate. A progressive enlargement of the area of the wing or airfoil would encourage the prolongation of leaps into short, controlled glides. In the minds of many, it is but an easy step from cursorial predation with occasional gliding to true flapping movements that might be needed to steer or correct instabilities. But all of this mathematical modeling assumes that Ostrom has correctly identified the wrist bones of theropods and that they are indeed homologous to those of *Archaeopteryx* and later birds. If the wrists aren't the same, then even if the mathematical modeling correctly predicts ways that a theropod hand might evolve, this model can tell us nothing about the actual evolution of avian flight.

The core issues of the debate are homology and function. Ostrom finds the wrists of dinosaurs, birds, and *Archaeopteryx* homologous; Martin does not. Further, even if they are homologous, Martin argues that *Deinonychus* could not bend its wrist laterally and cannot envision how a maniraptor-type wrist, with multiple planes of movement, could evolve into an avian wrist in which all bending except lateral bending is suppressed. For his part, Ostrom sees the beginnings of lateral bending in the wrists of theropods and *Archaeopteryx*, and can and does envision an evolutionary pathway that lies between the theropod wrist (as he interprets it) and the avian/*Archaeopteryx* condition (Figure 33). Although in talking with each scientist, I found myself swayed by the force of one or the other's logic and argument, emotional certainty is no basis for settling a scientific argument. What a particular scientist can or cannot imagine may motivate his or her research but does not constitute hard evidence. The only way to an objective resolution is to discover whether the bones in the wrists of birds, *Archaeopteryx*, and *Deinonychus* are or are not homologous—and the means are not yet at hand to do this. Empirical testing of proposed functions in a working, three-dimensional model of each wrist is another avenue for investigation, but in this case the complex nature of the joints and interactions involved is off-putting. A useful model would have to include representations of both the bones and the (unfossilized) cartilage, so a hefty dose of speculation would have to be swallowed at the outset. For now, the wrist debate must remain unsettled.

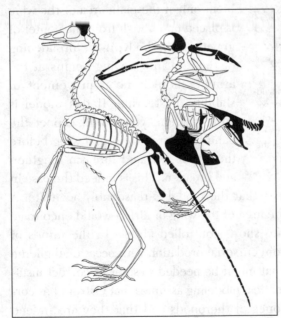

Figure 33. A comparison of *Archaeopteryx* (left) and a modern pigeon (right) highlights the similarities and differences. The areas of the skeleton where the differences are most striking (skull, sternum, rib, pelvis, and tail) are shown in bold.

Another contested point involves the finger structure of *Archaeopteryx* and whether it does or does not correspond to that in birds. Each digit (finger or toe) in a reptile, bird, or mammal is made up of a series of individual bones called phalanges. The first and second fingers on the hand of *Archaeopteryx* have, respectively, two and three phalanges. The question arises over the number of the phalanges in the third, outer digit of the hand. Although this region is preserved in several specimens, intense arguments raged over the significance of a disjunction in the middle of one phalanx in *Archaeopteryx's* outer digit. Starting in 1980, Samuel Tarsitano of Southwest Texas State University and his former supervisor, Max Hecht of the City University of New York, raised the possibility that the disjunction in the outer finger is a break, so that the bony elements on either side of it make up a single phalanx. That would make a total of three phalanges in that finger. Alternatively, in previous interpretations by Ostrom, Peter Wellnhofer, and others, this disjunction has been considered to be a joint, which would mean that the digit has four phalanges.

The difference in number of phalanges is significant because it implies which three fingers have been retained in *Archaeopteryx* from the primitive pentadactyl (five-fingered) condition. In various lineages of birds, reptiles, and mammals, one or more of the original five digits (numbered I–V) has been lost through the course of evolution; the question is, which ones have been retained? Currently, detailed embryological evidence tends to suggest that birds (or at least chickens) have retained digits II, III, and IV, as have lizards. In contrast, crocodilians have not lost any digits, but IV and V are smaller than the others. This observation has been taken as evidence that reptiles are more likely to reduce or lose digits IV and V. If this is true, then theropod dinosaurs (as reptiles) have also probably retained digits I, II, and III and cannot be ancestral to birds, which retain II, III, and IV.

Where does *Archaeopteryx* fall in this spectrum of finger choices? It depends how you count on its fingers. If *Archaeopteryx* has three phalanges (one broken) in its outer digit, then that digit is most probably digit IV. This would make *Archaeopteryx* like a bird in the hand (retaining II, III, and IV) and unlike a theropod (with I, II, and III): instant disproof of the dinosaur → bird hypothesis. But if *Archaeopteryx* has four phalanges in its outer digit, then that finger is III, making *Archaeopteryx* like a theropod but unlike a bird. After warm and lengthy debate, even Tarsitano and Hecht have conceded that the phalanx in question is probably not broken. In all probability, the outer digit on the hand of *Archaeopteryx* has four phalanges and is digit III.

While I was delighted to see some consensus approaching on *Archaeopteryx,* this resolution startled me, for it would seem to eliminate *Ar-*

chaeopteryx from being a bird or a bird ancestor. But there is a catch that turns the whole digital argument into an exercise in finger-picking. The ultimate sleight of hand in this argument is that no one actually knows that theropods retain digits I, II, and III. The entire dichotomy between avian and dinosaurian digital homologies on which the entire debate is based is illusory. There are so few dinosaur embryos that it is impossible to determine which digits are retained; indeed, this wasn't a simple or straightforward piece of research to carry out on living birds, for which an abundance of embryos at various stages of development is available. In short, the purported homologies of the wrist and hand of *Deinonychus* and *Archaeopteryx* are based on flatly equivocal evidence. These homologies are the cornerstone of Ostrom's theory of theropod ancestry—and the bedrock on which Martin and others have built their argument that something else (crocodilians, thecodonts, or pseudosuchians) must be a more viable bird ancestor than a dinosaur. The facts of the matter cannot now be resolved nor probably ever will be: we simply can't tell who is, or is not, a bird in the hand.

The tangle doesn't stop there. A similar problem pertains to the ankle joints of *Archaeopteryx*, birds, and dinosaurs. The issue is whether the bony object called the ascending process of the astragalus in dinosaurs, first pointed out by Huxley, is actually the same bone as is found in birds' ankles. Huxley found the two homologous. Martin and colleagues disagree, arguing that, in birds, this structure is more closely associated with a different ankle bone, the calcaneum, and not with the astragalus. In the hoatzin, a primitive South American bird, the bony process in question eventually fuses with the calcaneum and not the astragalus at all. Thus Martin calls this bony structure in birds the pretibial bone. To Martin, the pretibial bone is analogous (not homologous) to the ascending process of the astragalus in dinosaurs, for both function to strengthen the ankle joint in bipeds. That presents Martin with the problem of deciding where the pretibial bone came from in birds. His solution is to declare it a neomorph—a newly evolved structure in evolution—and one that is peculiar to the avian lineage.

This discussion spurred Chris McGowan, a paleontologist at the Royal Ontario Museum, to undertake some detailed research into avian embryology. In addition to hoping to resolve the issue of avian and dinosaurian ankle homologies, McGowan was intrigued by a much older question of diversification within the birds. Classically, ostriches have been seen as primitive and have often been compared to dinosaurs; one group of dinosaurs is even called the Ornithomimids, meaning the "bird mimics." If they are like dinosaurs, ostriches are not like "normal" birds. The group into which ostriches fall, the ratites, includes other large flightless birds like the emu, the rhea, and the extinct moas and elephant birds. They are fundamentally different from all other birds, which are classified as carinates. One obvious distinction is that all ratites are flightless. And although some carinates are also flightless, ratites have unusual anatomical adaptations to their flightlessness that unite them into a distinctive group. This fact inevitably leads to the question, Did ratites evolve via degeneration from flighted birds or did they diverge from the rest of the avian lineage before flight evolved?

McGowan thought a careful study of the formation of the ascending process or pretibial bone, whichever it might prove to be, in embryos of various ratites and carinates might clarify the issue. He used embryos of ostriches and emus as ratites and studied embryos of flighted quails and chickens to represent carinates. In addition to the embryo material, McGowan also examined X-rays of developing and juvenile ratites and carinates, too. He found that, as Martin had argued, carinates have a pretibial bone. It first appears in carinate embryos as a cartilaginous spur from the calcaneum, which is also still cartilaginous at that point. By forty days after hatching, young chickens have an ossified pretibial bone and an ossified calcaneum that have fused together into a single bone. If carinates represent all birds, then dinosaurs and birds have analogous and not homologous ankles. But McGowan also looked at ratites, and there the story was different. He found that an ostrich embryo close to hatching time has a prominent, well-ossified ascending process that is attached to neither the bony astragalus nor the bony calcaneum. By five weeks after hatching, this process has fused with the astragalus. Thus ratites show a primitive, dinosaur-like ankle, and carinates have a different and derived ankle arrangement. Huxley and Ostrom are correct that ratites and dinosaurs share this mode of growth and development of the ankle. Martin is also correct in that *some* birds (*viz.,* carinates) do not have an ascending process but rather a pretibial bone and therefore differ from theropod dinosaurs. There are two possible conclusions. Either ratites come from dinosaurs and carinates come from some other ancestral group—which seems inherently unlikely in view of the overwhelming number of detailed resemblances that unite ratites and carinates as birds—or ratites have retained a primitive, rather dinosaurian ankle arrangement, while carinates evolved a more derived ankle later. This research does not prove that dinosaurs are avian ancestors, but it does leave that hypothesis firmly in the "viable" category.

The final, debated point concerns the toes of these various species. Birds stand on their toes, which makes them digitigrade like dogs or antelopes rather than plantigrade (standing on the soles of their feet) like humans or bears. Birds also have three forward-facing toes and one backward-facing toe, known as the hallux. The hallux is the equivalent of our big toe. Because the avian hallux is rotated around to curve in the opposite direction to the other toes, birds can grasp objects (branches, prey, nesting materials) with their foot in much the same way that humans can grasp objects between their thumb and fingers. In birds, the hallux is opposed to the other toes (as our thumb is opposed to our other fingers) and is not simply set off laterally, like the big toes of great apes. This reversed arrangement of the big toe is called by paleontologists a "reflexed hallux" and is a classical feature of birds. Obviously, the reflexed hallux is essential for perching birds, but a reversed hallux, if not a long, grasping one, is also present in extremely terrestrial birds and even in flightless birds, like dodos. I couldn't help but wonder when and why the reflexed hallux evolved in the avian lineage.

The first step to answering these questions was to see if the reflexed hallux is already present in *Archaeopteryx*, and for once there seems to be unanimity. All of the toes of *Archaeopteryx* are clawed, even the hallux, which is much shorter than the other toes, and this is a birdlike condition. The hallux is also set at an angle very different from that of the other toes. John Ostrom interprets the hallux of *Archaeopteryx* as fully reflexed and so do several of his habitual opponents, including Larry Martin, Sam Tarsitano, and Alan Feduccia. Since *Archaeopteryx* already has a reflexed hallux, then the next logical step is to examine the conditions in the putative ancestors, to see if they represent a good starting point for evolving a fully reflexed hallux.

Theropod dinosaurs are the only major candidates for avian ancestry for which we have good specimens; no particular group of crocodylomorphs or thecodonts have been put forward as ancestors to birds, and so these groups cannot be evaluated. But theropod dinosaurs can be examined, and Ostrom believes that at least some of them had reflexed halluces. In this, he is following upon work by earlier scientists like Nopsca and Henry Fairfield Osborn, a paleontologist and onetime president of the American Museum of Natural History. The most unequivocal example cited by Ostrom is the foot of the skeleton of *Compsognathus,* which is nicely preserved in a position close to that in life (Figure 34). Ostrom sees the normal position of the hallux in this species as being at least partially reflexed, if not fully so, and virtually no one who specializes in dinosaur anatomy has disagreed with him on this point. But Martin and Tarsitano put a different construction upon the same information. They see the hallux of *Compsognathus* as being neither aligned with the other toes—a condition that would be distinctly *not* reflexed—nor fully reflexed in an avian manner; thus, they conclude, no dinosaur anyone can point to has a hallux that is reflexed in the avian manner. Their assessment may be accurate but less than telling. Like any other adaptive structure, a reflexed hallux is unlikely to develop in a single step. The question really is whether the theropod hallux deviates markedly from alignment with the other toes, giving it another plane of action that might represent the beginnings of a reflexed position; the answer to this question would appear to be "Yes, in some theropods." So if the stringent definition of "reflexed hallux" is "like a bird," then the theropod condition is excluded—but we can't expect ancestors to be as fully evolved as their descendants.

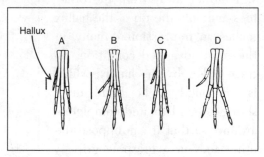

Figure 34. The skeletons of the feet of small theropod dinosaurs—*Coelophysis* (A), *Compsognathus* (C), and *Ornitholestes* (D)—show various stages of the shortening of the hallux and its separation from the other toes, compared to the reflexed condition in *Archaeopteryx* (B).

Another way of addressing the issue is to look at dinosaur footprints, as did Baron Nopsca early in this century. He was interested in understanding the anatomical adaptations to rapid running in dinosaurs and felt the bony anatomy could be usefully supplemented by the many dinosaur tracks found in Triassic beds. One of the key features for Nopsca was the arrangement and structure of the toes and feet. He could not find traces of an actual line of descent preserved as footprints, which would call for exceptional luck, but he constructed an illustrative evolutionary sequence out of tracks of different species that show the evolution of the position of the foot and, in particular, of the hallux, or big toe. Some forms, representing the most primitive condition, are fully plantigrade like a modern bear. Both the long toes and the elongated sole of such species are pressed against the ground.

131

The impression of the hallux is clear in the tracks of such species, and the hallux lies close to the other three toes. In other species' tracks, the impression of the hallux remains clear but it is somewhat rotated away from the other toes. In a third and more advanced group, the hallux is rotated still farther away from the other toes and only the tip of the hallux makes an impression. Finally, in the most advanced condition, no trace is left by the hallux, which is presumed to be both much shortened and rotated completely around, so that it can oppose the other toes in a grasping grip. A similar sequence representing different evolutionary stages in the evolution of the foot and hallux can be constructed from the tracks of digitigrade, or toe-standing, dinosaurs (Figure 35).

Nopsca cited this sequence in his works in the early twentieth century, and Ostrom and followers continue to discuss and figure the shortened and reflexed hallux of

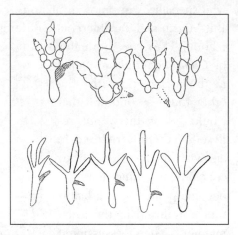

Figure 35. Baron Nopsca studied the footprints of dinosaurs to try to understand how the hallux or big toe evolved into the reflexed position. The top row shows the footprints of four Triassic dinosaurs that walked on their toes, as birds do; the bottom row shows a similar progression in Triassic dinosaurs that walk on the soles of their feet, as bears do. From right to left, the hallux moves progressively away from the other toes until, in the form on the farthest right, the hallux leaves no impression at all, presumably because it does not touch the ground.

small coelurosaurs. For Ostrom, the similarity of theropod foot structure to that of *Archaeopteryx* is a clear sign that bipedal theropod dinosaurs were already very birdlike in their hindlimbs. Theropods were, in a sense, preadapted to an avian way of life, whereas quadrupedal thecodonts or crocodylomorphs—neither of which has anything even resembling a reflexed hallux—were not. As Ostrom states the argument,

> [If] a quadrupedal flying stage had existed with avian history, it would be difficult to account for the total lack of hindlimb involvement in the flight apparatus of all modern (and apparently all fossil) birds. . . . The evidence of *Archaeopteryx* indicates that Nopsca was almost certainly correct in concluding that bipedal locomotion preceded the development of flight. In fact,

it seems most probable that even the initial stages of forelimb modification for flight did not take place until after an obligate bipedal condition had been achieved and the forelimbs had been completely freed from involvement in terrestrial locomotion and weight support.

Here is the crux of the issue: if a reflexed hallux is evidence of perching habits, like those of modern passerine birds, then the reflexed hallux did not need to appear until after flight had already evolved. The ancestor of birds thus could have been quadrupedal, like a thecodont or crocodylomorph, and arboreal, which would put it in a position to utilize the natural advantage of gravity in takeoff. On the other hand, if a reflexed hallux is evidence of a cursorial and predatory way of life and is present in avian ancestors, then these were both bipedal and terrestrial. The use of the reflexed hallux in perching would be, by this scenario, a secondary adaptation.

Because theropods have reflexed halluces in Ostrom's terms, the primary or initial function of this structure cannot be securing a stable position in an arboreal setting. He paints a gloriously mind-boggling picture:

> Advocates of the arboreal theory claim that the reversed hallux of *Archaeopteryx* is proof of arboreal habits. None of these authors mentions the fact that a fully or partially reflexed hallux is a characteristic of nearly all theropod dinosaurs . . . [and] none of these authors would advocate a tree-perching habit for *Allosaurus* or *Tyrannosaurus*. . . .

Indeed not. In fact, this image is so unacceptable that it undoubtedly fuels the fervor with which Feduccia, Martin, and Tarsitano deny that theropods have reflexed halluces.

But was a reflexed hallux initially an adaptation to perching or not? There are three features of the hallux that suggest its function. First, the hallux in both *Archaeopteryx* and theropods like *Compsognathus* is markedly shorter than the other toes (Figure 34). This disproportion in toe length is not generally found among living passerines or perching birds; it is more typical of ground-dwelling birds, like pheasants. A useful analogy here is the structure of the human hand and its uses. Apes have short thumbs that are not very effective in grasping because the disproportion in finger length makes it nearly impossible for them to oppose their thumb tips to their fingertips. In con-

trast, long-thumbed hu-
mans regularly use a precision grip—a thumb-tip-to-
fingertip hold—in manipu-
lating small objects. Even
the power grip that humans
use to wield larger objects,
such as a hammer, is facilitated by having a thumb
that is both long *and* opposable. Similarly, a short hallux would serve a bipedal
bird ancestor poorly in
grasping small objects, such
as berries or nesting materials, or in perching on
larger-bore branches. Thus,
if *Archaeopteryx* was a
perching animal, its feet are
not very well adapted to
that task in terms of hallu-
cial length. Does that mean
that the reflexed hallux is
primarily a predatory adaptation? Although predatory
ground-dwelling birds, like
the secretary bird or the
road runner, use their long

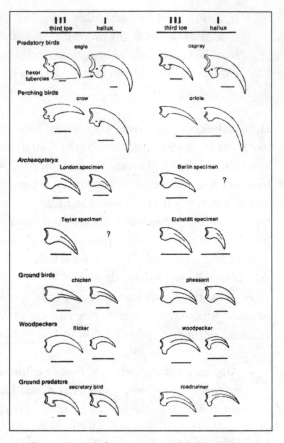

Figure 36. John Ostrom compared the bony claws
of *Archaeopteryx* with those of various types of birds.
Like ground-dwelling birds, *Archaeopteryx* has claws
that are robust and only slightly curved, and that have
only small tubercles for the attachment of the
muscle that flexes the claws.

reflexed halluces for grasping and holding prey with their feet, this function,
too, would be compromised by a short hallux.

The second indicator of the function of the hallux is based on the way in
which birds grasp with their feet. They contract muscles that insert on the
terminal phalanges—the last part of the toe—and the harder or more frequently those contractions are made, the larger the attachment site for the
muscle. That attachment site is the flexor tubercle, a bony bump at the
base of the terminal phalanx of the hallux, which in life underlies the horny
claw (Figure 36). Perching birds, who hold onto branches, and predatory
birds, who grasp struggling prey, have pronounced flexor tubercles that re-

flect a strong and habitual gripping action. In contrast, nonpredatory ground birds have flexor tubercles that are smaller and less distinct, because they have relatively infrequent reasons to grip strongly with their claws. In *Archaeopteryx*, the flexor tubercle is also small, much more similar to that of nonpredatory ground birds than to that of perchers or predators. Like its relative shortness, the weakness of the flexor tubercle suggests that the hallux of *Archaeopteryx* is poorly adapted to perching or gripping.

The third functional feature, first noticed by Ostrom also in 1974, is the curvature of the claws. Based on visual comparisons, Ostrom felt that the mild curvature of the claws in *Archaeopteryx* was also more congruent with terrestrial habits. Twenty years later, Alan Feduccia reexamined that assertion in a clever piece of research. Rather than eyeballing the relative curvature of the claws of different species, Feduccia quantified the geometry of the claws of modern birds to see how claw shape correlates with behaviors, choosing the horny claw of the central digit (III) as his focus. He documented the radius of curvature or claw arc in almost three hundred individual birds (Figure 37), selecting ten species to represent each of three categories (with about one hundred individuals in each): ground-dwelling birds, perching birds, and trunk-climbing birds. Although the three categories had distinctly different mean curvatures, there was enough variability that individuals from one category sometimes overlapped with individuals in an adjacent category. Ground-dwelling birds, such as the lyrebird, were the most distinctive, with relatively straight claws (with a mean curvature of 64.3 degrees), while perching birds like the bowerbird had more strongly curved claws (with a mean curvature of 116.3 degrees). Climbers, like various woodpeckers, had the most strongly curved claws of all (with a mean curvature of 148.7 degrees). Feduccia then measured the claw arc on the central foot digit of *Archaeopteryx*. While no one can prove whether this digit is embryological digit II or III, Feduccia assumed that *functionally* the central digits of

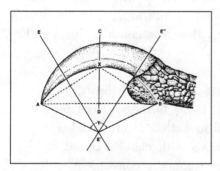

Figure 37. Alan Feduccia measured the curvature of the claws of *Archaeopteryx* and almost three hundred living bird species to determine the habits of the First Bird. His data suggested that the claws of *Archaeopteryx* had a curvature like those of perching birds today.

Archaeopteryx and modern birds were the same. In other words, if not homologous, these digits were at the very least analogous—and his question was, after all, functional, not phylogenetic. The central digits on the feet of the London, Berlin, and Eichstätt specimens have claw arcs of 125, 120, and 115 degrees, respectively, giving a mean value of 120 degrees. The claw arc of *Archaeopteryx* thus fell close to the mean value for perching birds, well above the mean for ground-dwelling birds, and well below the mean for climbing birds. This is a strong indication of arboreal and perching habits in *Archaeopteryx*, contradicting the conclusions based on the shortness of its hallux relative to the other toes and the weakness of its flexor tubercle.

Feduccia's fine study leaves two important comparisons unexplored—comparisons that might provide the key to resolving these opposing indications of function. As yet, we know a lot about the resemblances between the claws of *Archaeopteryx* and birds, but little about how closely the claws of *Archaeopteryx* resemble those of theropod dinosaurs. If theropods and *Archaeopteryx* also prove to have similar claw curvatures, then the bird data need rethinking. Maybe the interpretation that such a curvature indicates perching habits is incorrect—although it is certainly true that a modern bird with such curvature is a perching bird—or maybe theropods were not the cursorial, terrestrial species they are taken to be. Another issue raised by some skeptics is that Feduccia omitted particular types of birds from his study that might have changed the patterns of claw curvature that he observed. Since almost any study can be improved by using a larger sample, criticisms of sample size, when three hundred individuals have been measured, are generally nothing but carping. However, consider the birds Feduccia chose to omit from his study:

> . . . raptors (predatory birds), long-legged marsh birds, long-
> legged birds . . . that roost and nest in low bushes or trees, birds
> that resemble *Archaeopteryx*, and so forth . . . were avoided to
> eliminate as much as possible birds with claws adapted for
> strange habits or perceived to be generally convergent with
> those of *Archaeopteryx* for whatever reason.

Since Ostrom's hypothesis is that *Archaeopteryx* was a ground-dwelling predatory bird, information on the claws of predatory or otherwise similar birds might prove telling. But Feduccia felt that the claws of predatory birds, though curved, are too broad side-to-side to be similar to *Archaeopteryx;* unfortunately, he did not measure claw breadth systematically. Too, since *Archaeopteryx* was a long-legged bird, this fact might influence the biomechanical constraints on toe movements, so long-legged birds perhaps belong in the sample as well. Feduccia's study was well-executed and cleverly conceived; his results are so important, in fact, that they warrant expansion of the sample to include birds of additional habits and other aspects of claw shape.

The reflexed hallux of *Archaeopteryx* is also pertinent here, for a perching foot cannot perch securely without a hallux to oppose the other toes, whatever its claws are like. Feduccia sees the reflexed hallux as "strictly a perching adaptation . . . [that] would be a tremendous obstacle to running on the ground." By this, Feduccia probably means that a reflexed hallux as long as the other toes would be disadvantageous to terrestrial locomotion, for the hallux would be forever getting caught on things or causing the bird to stumble. However, *Archaeopteryx* didn't have a long reflexed hallux but a short one. As for the effect of a short reflexed hallux on terrestrial locomotion, it is logical to turn to the evidence of dinosaur footprints. Obviously, dinosaurs who made footprints in terrestrial sediments walked on the ground and, in at least some footprints, the hallux is clearly *not* aligned with the other, forward-facing toes and so must have been at least partially reflexed. In other cases, the hallux is so short that it has left no trace in the footprint at all. A short, reflexed hallux would seem to pose little hindrance to ground-dwelling. Because the hallux of *Archaeopteryx* is also short, I predict that—if we ever find its footprints—we will see only traces made by the three forward-facing toes and none from the hallux.

Despite the fervor of the arguments over the anatomy and function of *Archaeopteryx*'s body parts, a few conclusions are firmly supported and a number of others seem likely. *Archaeopteryx* is an animal that has largely or completely freed its forelimbs from a supporting, locomotory role. Its forelimbs are long, feathered, and tipped with exquisitely sharp claws that may be the only part of the hand not covered in feathers. Its wrist does not possess the adaptations for lateral bending involved in tucking the feathers up out of the way, nor does it have the shoulder adaptations that permit a wing flip to be made. Its wrist and hands resemble those of theropod dinosaurs closely enough for homology to be defended but not proven; the ambiguity of the situation is such that theropods, thecodonts, and crocodylomorphs all remain possible ancestors for *Archaeopteryx*. As for its hindlimbs, they are long, strong, and well-adapted to bipedalism. The ankle bears a locking or strengthening mechanism similar to that found in some modern birds and in some theropod dinosaurs. Clearly the hindlimbs carried most of the burden of supporting the body weight, except during whatever type of airborne locomotion was possible. The hallux was at least partially reflexed but so much shorter than the other toes that it would have been inefficient or perhaps even useless in perching. In any case, the hallux seems not to have been used in gripping strongly and regularly. And yet some of the evidence is contradictory. The curvature of the claws on the toes of *Archaeopteryx* is similar to that of perching or possibly trunk-climbing birds; the claw curvature is clearly unlike that found in ground-dwelling birds. Maybe there are other functions that require a similar claw curvature but only a short and relatively weak hallux; unfortunately, no one has identified them yet.

These limited conclusions are far from satisfactory. The anatomy is too ambiguous, the morphological gaps between birds, dinosaurs, thecodonts, and crocodylomorphs too broad to be spanned easily. Another clue to the locomotor abilities of *Archaeopteryx* is its feathers: what was their evolutionary function?

Chapter 6.

Birds of a Feather

———

Archaeopteryx had feathers and feathers are part of what make birds birds. All adult birds have feathers, or some degenerated vestige of them, whether they fly or not, and no other creature alive today does. These facts make a seemingly ironclad case for considering the feather to be an anatomical synonym for "bird." Feathers, combined with a reptile's toothy jaw and long bony tail, are why *Archaeopteryx* was recognized as a transitional bird-reptile in the first place. They are why *Archaeopteryx* is called a bird by almost everyone who knows anything about it. So perhaps it is surprising that there is still a fundamental question about the function of feathers: what are they *for?*

There are two main hypotheses, based on the attributes of modern birds. The first is the obvious, even self-evident one: feathers are for flying. Birds fly, birds have feathers, therefore feathers are for flying. I would accept this logic without question, except that, as an adolescent, I took great delight in constructing false syllogisms. The second hypothesis is that feathers are thermoregulatory devices—an avian version of fur—that were only later coopted for flight. Certainly feathers are critical in keeping birds warm, and I am firmly among those who feel that down-filled clothing is the only appropriate apparel for winter. So which function is primary, which secondary? As always, the condition in *Archaeopteryx* is key. The presence of feathers in the First Bird has been appreciated from the first discovery of the earliest specimen, but there is more to it than that. The structure and distribution of feathers, coupled with assessments of the aerial abilities of *Archaeopteryx,* provide a clear glimpse of the function of feathers. On the one hand, if feathers demonstrably predate flight—if the earliest feathered creature could not fly—then feathers cannot have evolved as flight mechanisms. Such a finding would not only clarify the function of feathers, but it would also imply that avian ancestors were already warm-blooded, as has been suggested of dinosaurs. On the other hand, if the earliest feathers found in the fossil record occur on an animal capable of true, flapping flight, then feathers are more strongly coupled to flight than ever.

The starting point of all discussion is the feather impressions on various specimens of *Archaeopteryx.* While these are readily recognizable, at least on the London and Berlin specimens, the exquisite delicacy of their preservation has actually worked against them. Some observers have found the impressions are *too* good, suspiciously good. There has twice been a flutter of accusations—the most serious occurring in the 1980s—that the feather impressions were in fact nineteenth-century forgeries, manufactured to serve the dual purposes of supporting the then-dubious theory of evolution and commanding a better price for the specimens. Specifically at issue are the first two skeletons of *Archaeopteryx,* the London and Berlin specimens that were found in 1861 and 1877 respectively. Each has feather impressions of unusual clarity; each first came to public notice while in the hands of the Häberlein family, who profited handsomely from the sale of these specimens of *Archaeopteryx.*

Fossil forgeries—those that have managed to fool specialists, at least temporarily—are rather rare, the prime example being the famous Piltdown skull of 1911–13. Although at the time of its discovery few expressed suspicion about this faked fossil, it was merely a clever composite of modern ape and human bones—the latter a few thousand years old—that were broken, stained, filed, and planted in a gravel pit where they could be conveniently "discovered" by an amateur paleontologist, Charles Dawson, who may in fact have been one of the perpetrators. Forty years later, once an impressive bulk of genuine fossils of early human ancestors had accumulated—and had repeatedly suggested a different anatomy from that observed in the Piltdown remains—the remains came under serious scrutiny and testing, which quickly revealed their falseness. Fossil forgers are not usually as talented as the Piltdown forgers, both in manufacturing the actual remains and in imparting to them convincing anatomical features that so neatly fitted the preconceptions of the day. Still, the nineteenth century was rich with various hoaxes and scams (stuffed mermaids, unicorns, and other interesting items)—so much so that some new and genuine animals, like the platypus, were initially treated with great skepticism. So it is not, perhaps, so surprising that the original *Archaeopteryx* find was rumored to be a fake, although scientists quickly rallied behind its obvious authenticity.

Those who raised the specter of forgery and falsehood in the 1980s were a disparate team, many based at University College in Cardiff, Wales: astrophysicists Sir Fred Hoyle and N. Chandra Wickramasinghe; Lee Spetner, an Israeli physicist and electronics expert; a physician, John Watkins; and another physicist, R. S. Watkins of University College, Cardiff. Hoyle and Wickramasinghe are well-known critics of the Darwinian theory of evolution, the latter having once quipped that the probabilities against "life having evolved by blind chance are about the same as against a whirlwind blowing through a scrapyard and assembling a perfect Boeing 747." From their perspective, a transitional bird-reptile such as *Archaeopteryx,* preserved in stone, is a literally incredible creature. All five were novices in palcontology when they leveled their accusations, which means they were unfamiliar with the ordinary effects of preservation and fossilization. But lack of expertise does not necessarily disqualify anyone from making acute observations. And, while naïve perspectives are sometimes warped, naïveté is also a powerful instrument for detecting flaws in well-accepted dogmas. The team of skeptics charged that some professional paleontologists had failed to notice the forgeries because they were blinded by the hallowed place accorded to the specimens. Other paleontologists had noticed the forgery, the skeptics claimed, but were more concerned with protecting the shaky theory of evolution (not to mention the credibility of the venerable institutions that housed the specimens) than with telling the truth. These accusations, if proven true, would be scandalous indeed. Although the accumulated evidence weighs heavily against the skeptics and in favor of the received wisdom that *Archaeopteryx*'s feathers are genuine, I found it entertaining and worthwhile to review the charges and their counterproofs.

Examining the original *Archaeopteryx* specimen in London, and photographs of it taken at low angles and high magnification, Hoyle and his colleagues noticed several suspicious circumstances. The feather impressions were made in a thin layer that seemed to them to be of an artificial and markedly different composition from the rest of the slab. Some of the apparently artificial material "looked under magnification like flattened blobs of chewing gum." On both the London and Berlin specimens, some of the feathers seemed to be double-struck—that is, they made one imprint, then apparently shifted slightly and made a second impression, which could not occur under normal geological conditions. The "double-struck" feathers had been noticed by Sir Gavin de Beer in his 1954 monograph on the London *Archaeopteryx,* but he never satisfactorily explained their presence. Hoyle and colleagues suggested a forger's error as the most probable explanation. Finally, they found that the two halves (the slab and counterslab) of the specimen were not perfect mirror images, suggesting less-than-skillful alteration. These accusations and observations pertained only to the Berlin and London specimens; the others were largely dismissed as "reptilian fossils . . . subsequently reclassified as *Archaeopteryx*" and were considered distinct from the first two finds, which "remain unique in that they are clearly of the same prototype and possessed unmistakable feather imprints."

The paleontological community was outraged by this attack on one of their most sacred objects and responded with vigorous protests and new studies. Geologists attested to the normalcy of the layer that contained the feather impressions and took umbrage at the statement that it looked like chewing gum. Of course the layer was fine-grained and indeed layered, they replied. Limestone is laid down slowly in very still waters and comprises layer upon layer of literally microscopic particles; these attributes are why the Solnhofen limestones have been sought for use as fine engraving stone for centuries. The features noticed by Hoyle and company were only to be expected.

Hoyle and his colleagues replied with a request for permission to take samples from the feathered and unfeathered portions of the original for testing. They were at first refused, a move that angered them greatly. Finally, Alan Charig of the Natural History Museum (then the British Museum [Natural History]) provided two samples to Hoyle and his team, who complained that the samples were "too small to make any chemical tests." They were able to examine the sample using high-resolution, scanning electron microscopy (SEM), which revealed some purportedly "foreign material." The British Museum dismissed the value of this observation, suggesting that the foreign material was preservative that had been applied to the sediment's surface to protect the fossil. The skeptics were embarrassed, insulted, and furious, writing:

> This reply also shows a somewhat devious attitude on the part of
> the Museum. They knew why we wanted the samples and they

knew what we intended to do with them. If they knew from the start that the surface of the fossil had been contaminated with preservative, why did they not give us a sample from material slightly under the surface?

The "testing" of the sediments speaks volumes about the naïveté of Hoyle and his colleagues on the subject of fossils. Many fossil specimens in museums—and virtually all rare ones—are routinely treated with preservative. It is standard protocol for a paleontologist intending to conduct SEM studies on fossils to ask the curators what the specimen has been treated with and whether or not it is acceptable to the curators that the preservative be removed with solvents. Hoyle and his colleagues had apparently never considered the possibility that preservatives had been applied. Since studying fossil surfaces under the SEM without removing preservatives is futile—analogous to trying to study the brushstrokes on a Rembrandt painting through a blanket—perhaps the museum curators simply expected that anyone requesting a sample for SEM inspection would remove the preservative. They may even have felt that offering unsolicited advice to so hostile a research team was far more than fair play required. And then, because Hoyle and his team found no evidence of forgery from their examination of the first two samples, the curators refused permission to take larger and more damaging samples. This decision might be subject to sinister interpretation but it is in fact standard procedure in many museums. A curator's first charge is to protect the specimens in his or her care. Requests for samples are frequently denied, especially if a researcher obtains ambiguous results from a preliminary study because it was poorly conducted. In retaliation, Hoyle and colleagues challenged the museum through the media to conduct specific tests of authenticity themselves. Being in a difficult position, the museum declined. If, as they expected, these tests yielded results supporting the authenticity of the specimens, then there were likely to be accusations that they had cheated or performed the tests incorrectly, given the distrust and suspicion that had already been engendered. Indeed, the refusal to conduct tests led Hoyle and colleagues to charge the museum openly with deliberate deception:

> We therefore think that they [the Natural History Museum] might have done the tests and are unwilling to report the results because they do not substantiate their contention that the fossil

is genuine. This hypothesis would also explain their adamant refusal to grant us further access to the fossil and it would also explain why they may be behind the move to silence us and prevent us from publishing our results.

We think that the Museum's position toward us gives cause for some suspicion that they may be hiding more than they are revealing.

The dismissive and ungracious attitude taken by the Natural History Museum and the paleontological community in this affair may seem unworthy and small-minded. On the other hand, the claims by Hoyle and colleagues lacked any real credibility and moved quickly from the realm of the irritating into the personally insulting. Neither side behaved in a way that would foster a civil exchange, to their mutual discredit.

Yet some of the concerns raised by the skeptics were addressed directly by the scientific community. To counter the scurrilous claim that the feather impressions on other specimens were dubious, new photographs were published of the Maxberg specimen, first found in 1955, that showed its feather impressions more convincingly. From an objective standpoint, the Maxberg feathers are much poorer in preservation than those on the Berlin and London specimens. This is perhaps not surprising since the Maxberg skeleton was decaying and disarticulated at the time it began to fossilize, as the bones show. Another counterargument is based on the infamous Teyler specimen, which was collected as a pterodactyl in 1855 and languished under this false identity until the feathers were seen by John Ostrom in 1970. How can the feather impressions on the Teyler specimen be explained away? Ostrom's sudden vision of feathers cannot have been fueled by wishful thinking or expectation, because his avowed wish and expectation was that he was examining a pterodactyl in the Teyler Museum. (There is no evidence that pterodactyls had feathers; in fact, skin impressions seem to indicate that pterodactyls had furry wings.) If the feather impressions on the Teyler *Archaeopteryx* are also the result of a forgery, then that forgery occurred four years *before* Darwin published *Origin of Species* and started the controversy over evolutionary theory. Even if a clever and knowledgeable forger had anticipated the imminent publication of such a theory and prepared forged material in advance, he or she was terribly careless to sell the Teyler specimen prematurely and cheaply—and then fail to call attention to those feather impressions, which remained unnoticed for more than one hundred years.

Another issue is the quality of the purported forgery. Siegfried Rietschel, director of the Staatliches Museum für Naturkunde in Karlsruhe, Germany, is the paleontologist who has conducted the most detailed study of the wing of *Archaeopteryx*. Rietschel tried repeatedly and failed to produce high-quality feather impressions using the wings and feathers of modern birds and various substances that were available in the nineteenth century. Because of the fine features visible in the *Archaeopteryx* fossils, real feathers are the most likely starting point for a forgery, but Rietschel simply could not discover a suitable method for impressing them into a stony substance. And, he observes, in order to make the tail feather impressions on *Archaeopteryx*, the purported forger had to conceive of and construct a new type of tail unknown in living birds and overlay it perfectly onto the extant reptilian tail vertebrae of the original fossil.

Obviously, any forgery must have been carried out after the slabs were split apart, revealing the genuine skeleton of a dinosaur fossilized within. A forger would need to put a negative impression on one slab and a corresponding positive impression on the counterslab; to be convincing, he or she would also need to align them perfectly. To Hoyle and his colleagues, the flaws in mirror-imaging in the slab and counterslab are the indelible fingerprint of forgery. But how different are they? Rietschel calculated that the differences between the slab and counterslab of the Berlin specimen occur in only about 5 percent of the total area of the specimen. This would leave the forger, if there were one, with the daunting task of producing excellent detail and near-perfect registration on an amazing 95 percent of the surface area of the slabs. Even this feat seems impossible, but another explanation for the imperfections must be found if forgery is to be discredited. Rietschel noticed that the flawed regions generally include those that were damaged during the original splitting into slabs or those in which the body cross-cut the plane of splitting. Either of these factors might account for the lack of perfect registration and mirror-imaging, without calling for extravagantly gifted forgers. There are also fine cracks within the specimen, difficult to see with the naked eye but readily revealed under ultraviolet light. These run through both the slab and the counterslab, through the feather impressions, and, in some places, actually through the fossilized bones. Such hairline cracks would be virtually impossible to manufacture if material had been added to the specimen.

Finally, Rietschel's microscopic studies of the plumage of *Archaeopteryx* also explain the double-struck feathers. He has shown that these are in fact the product of a double row of feathers, each of which has left its impression, rather than a single row impressed twice (Figure 38). Instead of six primary feathers—the number originally counted by Gavin de Beer in his 1954 monograph on the London specimen—*Archaeopteryx* had at least eleven. It is the imperfect or ghost images of the second row of feathers that give the appearance of a single, double-struck row.

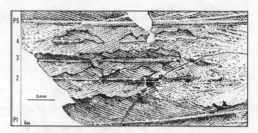

Figure 38. Siegfried Rietschel's study of the Berlin *Archaeopteryx* revealed that the "double-struck" feathers are actually two overlapping rows of feathers; when the top feather is broken away, the underlying one is revealed. Prior to this explanation, the existence of apparently "double-struck" feathers was one argument used by those who believed that *Archaeopteryx* was a hoax.

All of these data make the idea of adding something to the *Archaeopteryx* fossils seem extremely improbable, if not downright impossible. If the forger could not add a layer of material into which he had

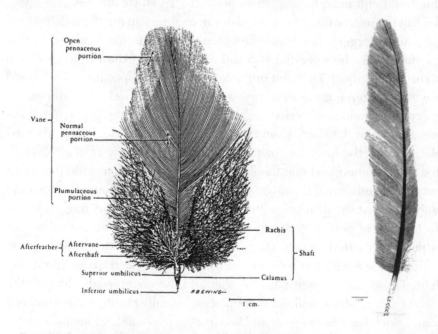

Figure 39. Both the contour feather (left) and the flight feather (right) of a modern bird have a rachis that divides the feather into two vanes made up of a series of barbs hooked to the adjacent barbs by barbules. This hooking mechanism allows the main part of the feather to act as a single, interlocking sheet, in contrast with the plumulaceous portion of the contour feather.

pressed modern feathers, could he or she have taken something away, by carving the feather impressions into the original rock? Here, too, the counterargument is compelling. In 1979, Alan Feduccia and Harrison Tordoff showed that the *Archaeopteryx* feathers are fully modern in their structure in virtually every respect (Figures 39 and 40). All feather impressions, as well as the original isolated feather of *Archaeopteryx*, are like modern feathers in that they have a strong central rachis that inserts into the skin at one end, which is called the calamus. The rachis divides the remainder of the feather into two vanes, each of which comprises a series of barbs: fine, elongated structures that branch from either side of the rachis and lie parallel to one another. Each barb is in turn attached to the barbs lying distal (toward the tip) or proximal (toward the calamus) to it by tiny hook-and-eye structures called barbules. The attachments formed by the barbules make the entire feather act as a single interlocking sheet rather than a series of separate hairs.

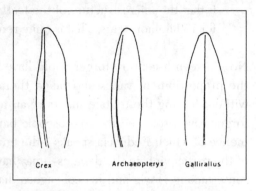

Figure 40. Alan Feduccia and Harrison Tordoff were the first to notice that the flight feathers of *Archaeopteryx* had asymmetric vanes, like those of a modern flighted bird (*Crex*). In flightless birds, like *Gallirailus*, the vane reverts to a central position similar to that in contour feathers.

Although this basic feather structure was readily apparent even at the time of the first discovery of an *Archaeopteryx* feather, Feduccia and Tordoff noticed something more. They found that the relative proportions of two vanes give an indication of the flight habits of the feather owner.

Feduccia and Tordoff saw that the flight feathers of *Archaeopteryx*, like those of modern birds, had asymmetric vanes, with the rachis placed closer to the leading edge. This arrangement makes the leading edge thicker in cross-section and stiffer, which means it is more effective as an airfoil. Among modern birds that are strong fliers, the leading edge vane is very much narrower than in poor fliers. And in modern flightless birds, the rachis reverts to a central, symmetric position such as is found in the contour feathers on the body (Figure 40). Thus *Archaeopteryx*'s feathers exhibit a peculiarity in their proportions that reflects an aerodynamic function.

The story of the publication of this paper is revealing, according to Feduccia.

The paper was actually rejected [initially]. One of the reviews said that the reviewer had seen all of the specimens of *Archaeopteryx* and the vanes were not asymmetrical. Well, you could almost look at the photos with a pair of binoculars and see that the feathers were asymmetrical. I began getting the feeling there was more to this than simple science. . . . People were vehemently arguing one side or the other without looking at the evidence. That's when I started getting into [the debate over *Archaeopteryx*]. The truth is that the microstructure of the feather is essentially unchanged for 150 million years. That is quite remarkable.

No nineteenth-century forger could have known about the importance of the proportions of vanes and faked them onto the Solnhofen limestone without leaving the telltale marks of an inscribing tool, nor could such a forger have produced the microscopic barbs and barbules. But the vehemence of which Feduccia speaks—the emotional urgency felt by so many of the participants in the debates—may have contributed to the developing sense among Hoyle and his colleagues that all was not on the up-and-up. Perhaps, not being ornithologists or paleontologists, they could not understand the tremendous importance of *Archaeopteryx* to practitioners of those sciences. And so perhaps Hoyle and his colleagues mistook the passion of the curators and experts for their view of *Archaeopteryx* and its abilities for the desperate fear of conspirators.

In short, the evidence for the purported forgery is flimsy, based on misunderstanding and misconstruction. Since the forgery charges were largely discredited, additional specimens have come to light that have settled the issue definitively. The Eichstätt specimen was reclassified as *Archaeopteryx* in 1988; its feather impressions actually pass under the bones of the wing, an extremely difficult task for a forger. Also, the Solnhofer Aktien-Verein specimen, found in 1992, boasts clear feather impressions. Both the Eichstätt and Solnhofer Aktien-Verein specimens have been prepared in the top museums of the world under scientific scrutiny, with little opening for clever forgery. The charges can surely now be laid to rest.

Nonetheless, these accusations of forgery sparked a round of studies that provide useful data for thinking about the initial function of feathers, which remains a vital problem. Feduccia and Tordoff end their paper about vane asymmetry with a bold statement of their conclusions about *Archaeopteryx* and feathers:

The shape and general proportions of the wing and wing feathers in *Archaeopteryx* are essentially like those of modern birds. The fact that the basic pattern and proportions of the modern avian wing were present in *Archaeopteryx* and have remained essentially unchanged for approximately 150 million years . . . and that the individual flight feathers showed the asymmetry characteristic of airfoils seems to show that *Archaeopteryx* had an aerodynamically designed wing and was capable of at least gliding. Any argument that *Archaeopteryx* was flightless must explain selection for asymmetry in the wing feathers in some context other than flight.

In a later paper, Feduccia makes some further points in support of his hypothesis about feather function:

Their aerodynamic design is attested to not only through structural analysis, but is proven by the fact that they lose their structural integrity and become relatively simple or hairlike when flightlessness evolves. . . . That body or contour feathers are designed identically to flight feathers and also become structurally degenerate in flightless birds is further testimony to the hypothesis that feathers evolved in an aerodynamic context. Once present, feathers were perfectly preadapted for serving the dual functions of flight and insulation in later endotherms.

In these and other works, Feduccia updates and rearticulates the feathers-as-flight-mechanisms thesis that was presented in detail in 1966 by Kenneth Parkes. The hypothesis supported by both Parkes and Feduccia rests on several points, the main one being the modernity of the feathers. Feathers are the most complex structure derived through evolutionary processes from skin: much more complex than hair or scales, for example. What, they might ask, is *Archaeopteryx* doing with a feather structure that is demonstrably shaped by aerodynamic needs if it has no aerodynamic needs? The case is complicated by the presence of a stout furcula or wishbone in *Archaeopteryx*, long taken as a clear adaptation for flight. As the work by Jenkins, Dial, and others showed, in birds the furcula is a sort of spring-loaded spacer that acts both to brace the shoulders outward and to vibrate in and out with each flap. Yet there are growing numbers of reports of furculas in dinosaurs that are clearly not airborne, like *Allosaurus*. Here, then, is a second structure that today functions exquisitely well in a flying creature and is also found among nonfliers. Does this mean the furcula had one function in terrestrial dinosaurs and was coopted for flight in the avian lineage? Structures do change function over the course of evolution, or else no creature could ever evolve to do something that its ancestors did not do without developing completely novel structures—neomorphies—which is considered to be a rare occurrence. Or does this mean that the V-shaped structures recognized as furculas in birds and *Archaeopteryx*, on the one hand, and dinosaurs, on the other, are two different structures? The issue cannot be resolved at this point, but it is troublesome.

Only in having feathers and a furcula does the anatomical evidence suggest *Archaeopteryx* was flighted. Virtually every other adaptation to flight that has yet to be enunciated is absent in *Archaeopteryx*. Parkes and Feduccia freely admit that *Archaeopteryx*, as a flier, suffered from such handicaps as its lack of a keeled sternum; the weakly developed bones of the shoulder and upper arm, which in modern birds are much more robust because they anchor the massive pectoral muscles that power flight; its lack of a triossial canal, which is needed to perform a wing flip; and its segmented bony tail, in distinct contrast to the pygostyle, which helps the tail feathers of modern birds act as an effective rudder for steering and control of pitch and yaw. The joints of *Archaeopteryx*'s forelimb do not form a mechanical linkage system comparable to the one in living birds that makes opening and closing the wing so energetically effective. Parkes even doubts that the elbow joint of *Archaeopteryx* was strong enough to withstand the stresses of a powerful downstroke at all.

How can the adaptations of feather and furcula be reconciled with the contradictory evidence of the arm, wing, and tail structure of *Archaeopteryx*? Would flapping flight or some other type of gliding have been possible with so few of the anatomical adaptations seen in modern fliers? Feduccia thinks the answer is yes. Writing with Storrs Olson, an avian expert at the Smithsonian Institution, Feduccia demonstrated that the furcula found in *Archaeopteryx* is large, even hypertrophied. In modern birds, the large pectoral musculature that powers the flight stroke comprises about 15–30 percent of the total body weight. These muscles are attached to the humerus or upper arm bone on one end, and arise from the furcula and the bony, keeled sternum on the other. Feduccia and Olson hypothesize that the condition in *Archaeopteryx*—which has a smallish humerus, no keeled sternum, and an extra-large furcula—represents a first step toward the modern condition, sufficient to have supported a well-developed pectoralis muscle, if not one as proportionately large as in modern birds. They believe the extra-large furcula afforded ample area for attachment of a set of muscles that could have performed a flight stroke, even without a bony carina on the sternum. According to the experiment by Sy described earlier, only pectoral muscles and not the supracoracoideus that performs the wing flip are necessary for horizontal flight. If a bird with a severed supracoracoideus tendon can fly horizontally, then so (presumably) could *Archaeopteryx* with pectoral muscles as reconstructed by Feduccia and Olson. However, they suggest that the absence of a triossial canal in *Archaeopteryx* meant that a wing flip and takeoff from level ground were impossible. They propose that the carina, or keeled sternum, evolved later, along with the triossial canal, which were anatomical refinements that enhanced aerial performance.

Feduccia's contention that vane asymmetry is indicative of aerodynamic capabilities has been weakened by additional research conducted by J. R. Speakman and S. C. Thomson from the University of Aberdeen, who investigated the variation in vane asymmetry that occurs within the wing of a single individual. In addition to varying with flight capabilities, as noticed by Feduccia and Tordoff, the degree of vane asymmetry depends upon the area along the feather at which the vanes are measured and upon the position of the feather within the wing, according to Speakman and Thomson. Measuring feathers from seventeen species of flying birds, they showed that the greatest degree of asymmetry occurred at a point about 25 percent from the feather tip. At this 25 percent point, the first four primaries showed the highest degree of asymmetry, which was markedly diminished in other primary or secondary feathers. They divided the width of the trailing vane by the width of the leading vane to derive an asymmetry ratio and, with data from a larger sample, confirmed the locomotor effect observed by Feduccia and Tordoff. Feathers from seventy-eight birds that used flapping or soaring flight showed high degrees of vane asymmetry in the first and fourth primaries at the 25 percent point. Asymmetry ratios for first and fourth primaries ranged from 2.22 to 11.75 with an average close to 6. There was no evidence of a significant decline in asymmetry ratio between the first and fourth primary in flighted birds, but asymmetry ratios fell below 4 in primaries higher than the fourth or in any secondary feather. A sample of first and fourth primaries from eighteen flightless species, measured similarly, also had significantly lower asymmetry ratios that ranged from about 0.75 to 4.25.

Where does *Archaeopteryx* lie? The isolated feather of *Archaeopteryx* yielded an asymmetry ratio of 2.2 at the 25 percent point; however, there is no way of knowing which feather this specimen represents. If it is one of the first four primaries, then its asymmetry ratio lies just at the bottom of the range for flighted birds and within the range for flightless birds. If it is a later primary, then its earlier primaries were probably more strongly asymmetrical, suggesting flight capabilities. Speakman and Thomson were also able to measure the asymmetry ratios on the fourth, fifth, and sixth primaries of the Berlin specimen and the third and fourth primaries of the London specimen, which averaged about 1.45. The average ratio from *Archaeopteryx*'s fourth primaries was lower than the lowest ratio (2.2) in any of the flying species in their sample and is statistically the same as the asymmetry ratios for the first primaries of flightless birds in their sample. Speakman and Thomson observe that their results are consistent with suggestions that *Archaeopteryx* was incapable of flight.

Their data certainly seem to undercut Feduccia's ideas, but Feduccia is not swayed. He identifies the isolated feather as a secondary, and not a primary, feather. He says, "You have asymmetry going all the way into the secondary feathers. *Archaeopteryx* has to be a strong-flying bird; it makes no sense otherwise that even its secondary feathers are asymmetric." He also returns to the fact of the complexity of feather structure and the superb way in which a feather is suited for flight—points which, as an expert on feathers, he weights heavily.

> Everything about feathers indicates an aerodynamic function. They are lightweight, they are excellent airfoils, they produce high lift at low speeds, they have a Velcro-like quality that lets them be reassembled if they have been disrupted. . . . Feathers have an almost magical construction which is all aerodynamic. Now, of course, they are also insulatory, but how do you evolve all those features for insulation initially? Instead of using hair, a simple integumentary derivative, you're going to use the most complex structure . . . I don't think evolution works that way. It would be gross evolutionary overkill to produce feathers for insulating a hot-blooded dinosaur.

Underlying all of this research is an important but unspoken assumption: that vane asymmetry has been stable through the course of the evolution of flight. It is impossible to determine whether this assumption is correct. Although vane asymmetry obviously has mechanical effects on the primaries that make flight easier and more efficient, earlier forays into flight may not have involved this adaptation—or may have involved this adaptation to a less pronounced degree. Certainly, the feathers of *Archaeopteryx* show some degree of asymmetry and, in the case of the isolated feather—no matter where in the wing it was placed—the asymmetry is fully comparable to that seen in living, flighted birds.

Central to all theories about the initial function of feathers are beliefs about the origin of feathers. It is universally accepted that feathers evolved from scales, whether the ancestor in question was a dinosaur, a thecodont, or a pseudosuchian. Like hair or scales, feathers are a specialization of the epidermis, the outer layer of skin, that characterizes a particular group of animals. But I can't help asking: why did feathers develop at all? In seeking an evolutionary explanation, researchers have tried to identify a selective advantage that would fuel the evolution of feathers from scales. Heilmann, Savile, and others have shown that even a slight elongation of the scales along the trailing edge of the forearm, somewhat like the suede fringe on a cowboy jacket, would give an advantage to an arboreal animal at risk of falling out of the trees. But, unlike the swinging suede strips on the stylish jacket, the scale fringe would benefit by being stiff, for length and stiffness are properties that would make the fringe maximally effective in transforming plummeting to controlled falling or parachuting. Progressively elongating and elaborating scales into proto-feathers and then true feathers would permit controlled gliding and then flapping flight.

Do the feathers of *Archaeopteryx* support this scenario? The answer depends on which type of feather evolved first, for birds have pennaceous and plumulaceous feathers that serve different purposes. Flight feathers and body contour feathers are classified as pennaceous feathers; they have the strong central rachis and vanes comprising barbs and interlocking barbules. Near the bottom of a pennaceous feather, where it inserts into the skin, the barbule system deteriorates, and there is a small, loosely linked plumulaceous portion that looks hairier and less feathery. Pennaceous feathers are what *Archaeopteryx* has, and these are the type of feather that would be most useful in the early stages of aerial locomotion. Down or plumulaceous feathers would be of little use in flying and are believed by some scientists to have developed secondarily from pennaceous feathers. I can easily imagine the process of degeneration that would transform a pennaceous feather into down: thinning the rachis would make it very flexible, not stiff and strong; removing the barbs and barbules would enlarge the plumulaceous portion until it extended over the entire feather. A similar sort of degeneration of flight feathers occurs in living flightless birds that are descended from flighted ancestors. In this scenario, pennaceous feathers evolved for flight in proto-avians or very early birds. At some later point, the addition of warm-bloodedness would offer a further advantage by prolonging the hours and temperatures at which flight could be used for foraging. But warm-bloodedness means sometimes keeping ambient cold out and heat—generated by both metabolic and muscular activities—in and shedding excessive heat at other times of day or year. Down, with its fluffy structure and abundant air spaces, does a very good job of preventing heat exchange between the body and the environment. It is found mostly on young birds that cannot yet fly, chicks more in danger of becoming cold than of accumulating excessive heat. Down appears to be a simple evolutionary modification of pennaceous feathers, rather than a novel evolutionary structure, implying that pennaceous feathers preexisted the need for thermoregulation. If the scenario is played out the other way—down first—and thermoregulation is the initial need to be met, then another type of insulator might have evolved instead of a feather. As Parkes puts it, "Why not hair?"

This scenario of feather evolution is logical and tidy, but there is precious little direct and independent evidence to either support or refute this reading of the facts. Surprisingly, facts and observations cited by Feduccia or Parkes as evidence that pennaceous feathers are the primary type are turned on their head by Philip Regal, to make them evidence of the primacy of thermoregulation in avian evolution.

For example, Regal completely discounts one line of argument used by Feduccia:

> His main claim is that whenever birds become flightless they lose the pennaceous structure of the feathers. But his interesting observation might suggest . . . only that the pennaceous structure is today maintained by aerodynamic selection pressures and it says nothing of origins. . . . The loss of pennaceous structure in flightless birds may show that for an endothermic animal where the problem is largely heat retention, hair-like structures [i.e., down] are indeed about as useful as the more complicated pennaceous structures.

Whereas Feduccia points to the wonderful structural complexity of feathers as evidence of selection for aerodynamic properties from the outset, Regal is incredulous that such a complex structure evolved just for flight. Pennaceous feathers are excellent devices for flight, and plumulaceous or degenerated feathers are useless for flight. But what adaptive advantage to aerial locomotion can be imagined for the transitional stages between scales and fully modern pennaceous feathers? None, in Regal's view. Imagine a half-pennaceous feather—say, one with fine barbs but no interlocking barbules, or one with grosser subdivisions than the tiny barbs of modern birds. Neither would function well in an aerodynamic setting, if at all. Thus, while Parkes may well ask, "Why not hair?" if the problem were thermoregulation, Regal counters with "Why not wings made of skin?" if the problem were to evolve an airfoil. He explains,

> Birds are basically flying animals and they are unique in possessing feathers. . . . Animals that fly or glide—flying lizards, flying squirrels, flying fish, bats, pterosaurs, and the vast majority of flying insects—have broad continuous surfaces as parachutes or airfoils. Birds are unique among vertebrates in having that portion of the wing that provides lift comprised of many individual units and in having a separate portion of the wing that provides thrust . . . made up of a number of elements.

In every other flier, the portion of the wing that provides thrust is a solid sheet. Birds have remiges (individual flight feathers) that act together as a unit but are also capable of separating into a slotted structure to facilitate stalling out when that movement is needed. While he accepts the inherent logic of the scenario of a fringe on the forelimb becoming progressively elongated to help control falling, Regal sees this selective trajectory as occurring only after feathers have been formed. "This line of reasoning," he argues, "explains only the evolution of wings . . . and *not* of feathers."

Regal's alternative hypothesis is that primitive feathers are a simple but important advance over scales in terms of thermoregulation. And, Regal observes acutely, keeping heat out can be just as important to survival as keeping heat in. Flying is the sort of activity that generates a large amount of metabolic heat. If this problem is to be considered, then one important consideration is *when* that flying occurred. The likelihood is that the original bird and its ancestors were diurnal creatures, active primarily during the day. Certainly most birds are diurnal today and only a few lineages—primarily the owls and their relatives—have specialized in working the night shift. This premise of diurnality is supported by the skull of *Archaeopteryx*, whose orbits do not show the extreme enlargement of a nocturnal species.

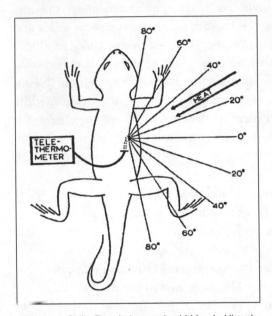

But if all of the early bird's flying is carried out during the daytime, the problem of heat is intensified. The outspread wings increase the surface area exposed to the sun's rays, so wings add additional heat to the system, already challenged by the vigorous activity of flying itself. On the one hand, convective heat loss can compensate for some of the metabolic heat—especially *if* the wing is uninsulated (like a bat's nearly naked skin wing) and *if* the flights are performed at times of day when the ambient temperature is lower than the internal temperature of the flier. On the other hand, in open areas on warm, sunny days, absorption of radiant

Figure 41. Philip Regal observed cold-blooded lizards manipulating their body temperature by changing their position relative to a heat lamp. When the rays were directed perpendicular to the long axis of the lizard's body, the scales acted like tiny parasols and shielded the lizard from absorbing more heat.

heat during flight would pose a potential problem *unless* the wing were covered in a substance that could be used to retain or repel heat as needed. Do feathers fit the bill?

Regal was led to wonder about such questions not by the study of birds per se but by watching lizards with enlarged scales in the laboratory. Of course, avian feathers did not evolve directly from lizard scales, but they did almost certainly evolve from some type of scale, so many of the issues are the same. Through his experiments, Regal found that lizards were able to fine-tune their body temperatures simply by changing their position relative to the heat lamp. If cold, they exposed maximum skin surface to the heat lamps as they basked, letting the warming rays penetrate to their skin; once heated, they altered their position so that the scales functioned as tiny parasols, shielding their skin from further warmth. Measurements of the rise in skin temperature of lizards in different positions showed that the orientation of the scales (and thus the body) relative to the heat lamp significantly changed the amount of heat absorbed (Figure 41). The most heat was absorbed if the heat lamp was positioned at the animal's tail and directed forward onto its body; somewhat less was absorbed with the lamp at the animal's head; and the least heat was absorbed when the beam was perpendicular to the lizard's long axis. The difference lay in the cumulative exposure to the heat lamp, which was determined by the size of the tiny shadows cast by each scale. In the wild, basking lizards actually make these adjustments to their position in response to their thermoregulatory needs. Regal then produced models of scales out of paper and was able to show experimentally that there were meaningful differences in the effectiveness of their heat shielding ability that related to the size of the scales. Larger or longer scales were the best heat shields, suggesting a selective advantage might accrue to elongating scales into proto-feathers. The longer the proto-feather, the greater its potential range of responses to climatic conditions.

Feathers, even proto-feathers, are not scales, of course. Scales are living tissue, capable of transferring heat to or from the body via the blood supply, in the same way that a membranous patagium, or skin wing, does. Feathers are dead tissue with no potential for direct heat transmission. They cannot be sunburned nor can they directly absorb or dissipate heat. However, un-

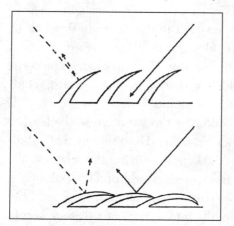

like a lizard's scales, a bird's feathers can be raised or lowered, which enhances their potential as thermoregulatory devices enormously (Figure 42). For example, in cold weather, birds can fluff their feathers up, enlarging the net size of the insulating air pockets between the feathers and helping them to retain warmth. Birds that are cold or wet may also extend their wings to increase the surface area exposed to the sun, raise their feathers, and may turn so that the sun's rays can penetrate to the skin exposed between the feathers. Lowering the feathers helps exclude excess radiant heat and protects the thin skin from sunburn.

Figure 42. Because feathers are mobile, they are even more effective at regulating body temperature than a lizard's fixed scales. Depending upon the angle of the sun's rays (indicated by arrows) relative to the bird's body, feathers can be raised or lowered either to permit the heat to penetrate to the skin or to protect the skin from further heat absorption.

The ability to raise or lower feathers and thus control heat loss or gain is particularly important to birds because they—like lizards—lack sweat glands. Birds, mammals, and lizards can pant to lose heat, but this option is not enormously effective under extreme conditions of high activity and high ambient temperature. Thus evolving an additional and highly effective mechanism for losing heat could be crucial to the success of adopting an extremely strenuous lifestyle, such as one in which flying is the main means of moving around to feed. But flight does not necessitate warm-bloodedness, and Regal explicitly rejects the presence of feathers on *Archaeopteryx* as de facto evidence of warm-bloodedness. As he points out, both endothermic and ectothermic animals face thermoregulatory challenges that, if solved, prolong either their daily hours of activity or their seasonal activities. An ectothermic species, relying primarily on basking to raise its body temperature, would gain a selective advantage by evolving a more effective means of regulating its temperature. Movable proto-feathers might be useful either for enhanced basking or for enhanced heat loss.

Of course, hair is another evolutionary solution to the thermoregulatory problem: a popular one adopted by all mammals. Fur or hair keeps warm-blooded mammals warm, although some still bask in the sun or huddle together in the cold to save energy. Fur also helps mammals to stay cool by shading their skin from the sun. In 1958, an inventive researcher named R. B. Cowles performed a wonderful experiment to demonstrate that hair or fur keeps heat out as well as in. He made a fur coat for a lizard and then tested the lizard's tolerance for exposure to heat. The furry lizard could endure exposure to heat for much longer than the naked lizard, cold-bloodedness and low metabolic rates notwithstanding.

Regal also offers a different explanation for the structure of the feather that owes nothing to the use of feathers in flying. He observes that subdividing scales or proto-feathers—the first step in evolving the anatomically complex structure of modern feathers—provides additional flexibility. A simple subdivision of the scale into two parts, then four, and so on until the scale had been transformed evolutionarily into a multipart feather, would constitute progressive improvements in flexibility. At each stage, there would be an advantage to simple interlocking mechanisms like proto-barbs, which could be arranged along the edge of one part and attach it to the next adjacent part, for they would impart continuity without sacrificing flexibility. Evolving a tough but elastic covering offers great advantages. A more flexible covering is less prone to damage than a rigid one; additional protection is offered by having an interlocking, sheetlike structure, so that blows from vegetation or predators are dissipated across a broad area. This combination enables birds to have thin skin because they also have exquisitely effective feather "armor" to protect that skin. In short, Regal finds feathers wonderfully adapted to perform a protective and thermoregulatory role during early avian or preavian evolution. Regal's scenario accounts for a selective advantage both to forming feathers from scales and to evolving an intricate, interlocking microstructure within feathers, with accretional advantages at each step. Thus he concludes that feathers may have been designed for thermoregulation, which would preadapt them for both endothermy and flight.

Both theories—feathers-as-flight-mechanisms and feathers-as-insulation—have a good logical basis that renders them plausible. Neither can offer much support in terms of hard evidence. Practically every point in each argument can be inverted and used by the opposition. Other lineages of vertebrates have evolved other flying mechanisms *and* other thermoregulatory devices, demonstrating the importance of each but the primacy of neither. It is an unsatisfactory state of affairs. But there is one more aspect of feathers and their evolution yet to be considered. This is their collective entity, the wing. Does the function of the wing reveal anything about the origin of feathers?

Chapter 7.

On the Wing

Ask almost anyone what wings are for and the answer will be "flying." When the subject is birds, it is practically impossible to consider the function of wings without becoming entangled in the question of the function of feathers, and there is no simple route out of that maze. But when I ask biologists who examine wing function from the perspective of insects, I get a different answer. Although insects are only very distantly related to birds—and although the wings of insects and birds are made up of very different substances—the insights gleaned by insect specialists into the evolution and function of wings are relevant.

The taxonomic gulf between insects and birds is enormous. No one would suggest that insects and birds followed the same evolutionary route in evolving flight. Still, the functions, problems, and solutions that span this great gulf are likely to pertain to fundamental truths. Flapping flight is neither a simple mode of locomotion to evolve nor an easy one to comprehend, whether hummingbirds or dragonflies are the subject. However, the catch in understanding the evolutionary appearance of a key innovation, like a seeing eye or a flying wing, has always been the apparent uselessness of the transitional stages.

"What good is half an eye, or the nub of a wing?" ask experimental biologists Joel Kingsolver of the University of Washington and Mimi Koehl of the University of California at Berkeley, echoing a traditional bewilderment. Darwin tried to answer this and similar criticisms of his theory of natural selection by suggesting that the function of an organ may change through time. Thus, despite an apparent structural continuity of the wing (or eye) as it evolved, the structure performed a series of different functions through time. A wing might not perhaps have been "about flying" from the outset, but only later in its evolutionary history. The essence of evolution, as always, is that things change.

Mimi Koehl is a nonconformist biologist, one who happily complements the heavy-duty theorizing that most professors deem appropriate for discussions of major evolutionary issues with direct experiments designed to answer her questions. Her experiments are sometimes faintly cockeyed but always interesting. She has, for example, built model insects and flown them and tethered a living copepod, a type of zooplankton, by supergluing a leash made of hair to its thorax so that she could film its feeding movements through a microscopic lens. Koehl is also an engaging teacher, perhaps because her enthusiasm for her subject is so apparent. In lecture, she switches freely from impromptu imitations of the invertebrates she studies to highly technical discussions of the physics of movement and locomotion. Joel Kingsolver, once her postdoctoral student and now a professor at the University of Washington, often collaborates with Koehl; one of their best-known projects focused on the origin and function of wings.

Their starting position in this research is that they find Darwin's argument for functional change cogent, but unproven. Perhaps wings evolved for a reason other than flight, they agreed, but any attempt to trace the evolutionary development of insect wings is hindered by the deficiencies of the fossil record of insects. "There is no insect *Archaeopteryx*," they lament in one of their publications. The irony is that the excellence of *Archaeopteryx* specimens hasn't produced any conspicuous agreement over the function of flight structures or the evolution of flight in the avian lineage anyway.

The enigma tackled by Kingsolver and Koehl has strong parallels to that over the function of feathers in *Archaeopteryx*. As they summarize the situation, the competing hypotheses for the selective factors involved in the origin of insect flight are numerous and not necessarily mutually exclusive; they cycle regularly in and out of vogue, but cannot be resolved by evidence from the fossil record as it currently exists. The hypothetical ancestor of flying insects is the protopterygote—pterygotes are flying insects—a hypothetical ancestor directly analogous in position to "Proavis." Rather than add up the evidence about the features of the protopterygote, Kingsolver and Koehl turn to an approach they call "bounded ignorance." They explain:

> Rather than attempt to identify a profile for the most likely pro-topterygote, we describe a range of possible morphologies that seem consistent with available evidence. Instead of constructing a specific scenario for the selective factors involved in the evolution of flight, we explore a variety of plausible functions of which we ask: Do small changes in wing morphology [shape] result in functional changes (that is, would selection on wing morphology result)?

They identify three critical junctures in the evolution of insect wings. First, the initial proto-wing must offer a selective advantage over the wingless condition; second, increases in size of the proto-wing must offer a continuing selective advantage by improving performance; third, developing a movable proto-wing must also impart an additional selective advantage. The first and last critical junctures do not directly parallel the evolution of avian flight. Although avian ancestors were technically wingless, they evolved wings by modifying extant thoracic appendages (forelimbs). Thus the issue in avian evolution is one of increasing the size of a preexisting structure, not of developing a novel structure. And, since avian wings arose from a forelimb that was intrinsically movable, I find it hard to imagine how birds or their ancestors could ever have had an immovable wing. In contrast, insects kept all of their legs and added a new appendage, a novel structure, when they first evolved wings. Quite plausibly, this appendage started either as an aquatic gill or as a rigid outgrowth of the exoskeleton. So if there is insight into the origin of wings to be transferred from insects to birds, it will apply only to the effects of increasing the size of an already-extant wing.

Two of the main hypotheses about the origin of insect wings are also very similar to those in the debate over bird flight: one is that insect wings developed as thermoregulatory devices and another is that they evolved to control falls or leaps, eventually leading to parachuting and gliding en route to flapping flight. These hypotheses are at least in theory able to be refuted or supported experimentally. More difficult to evaluate is the third hypothesis: that insect wings evolved as elements in courtship display. Wings are used in sexual displays by modern insects, and secondarily flightless insects that have lost their hindwings may retain their forewings so that they can still "sing" to attract females. A similar idea about theropod evolution has been proffered by Gregory Paul, who suggested that feathers may have evolved in dinosaurs as part of a visual communication system that would be used in courtship and other social interactions. If more effective courtship displays—i.e., bigger wings—mean that a male gets to impregnate more females, then sexual selection might act rapidly to increase wing size. The hypothesis linking wings to display is thus logical but untestable at present. The fourth theory (which doesn't apply to birds) is that insect wings evolved as gas exchange devices used by the aquatic nymphs of primitive insects.

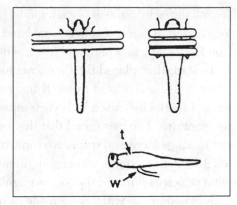

Figure 43. Mimi Koehl and Joel Kingsolver examined the effectiveness of insect wings as thermoregulatory and/or aerodynamic devices, using the simple models shown here. A wing of any size can affect the thermoregulation of an insect, but only wings larger than 20 percent of total body length had an appreciable effect on the aerodynamic properties of the models.

To test the effects of getting wings and enlarging wings on thermoregulation and aerodynamics was relatively simple. Koehl and Kingsolver made various physical models in the shapes of fossil protopterygotes (slim cylindrical bodies or wide, flattened bodies) with heads, legs, and zero, one, two, or three pairs of wings attached to their thoracic segments or abdomen (Figure 43). These models ranged in size from two to ten centimeters in length and bore a good general resemblance to insects like dragonflies (which are preserved in the Solnhofen limestones) (Figure 44). They made the model bodies out of epoxy, which has a density and thermal conductivity similar to those of present-day insects. The legs and wings of each model were made of different substances according to the properties being tested. To look at temperature changes, Koehl and Kingsolver constructed legs out of wire and wings out of either construction paper (which has low conductivity) or aluminum foil (which has high conductivity); they also embedded a thermocouple wire in the

Figure 44. The preservation of the Solnhofen fossils is so exquisite that even delicate creatures, like this dragonfly, are readily recognizable.

second thoracic segment of each model to record changes in temperature. To look at aerodynamic abilities, Koehl and Kingsolver made models with balsa wood legs and wings of thin plastic membrane enclosed by copper wire.

To start, they placed the thermoregulatory model in a wind tunnel, simulating flight, and waited to see if body temperature changed. It did. Every winged model increased its body temperature to more than the ambient air temperature. The pair found that the total heat gain increased progressively as the wings increased from zero (no wings) to an optimal one centimeter in length. At this point, neither lengthening the wings further nor changing wind velocity improved the thermoregulatory function of the wings. Surprisingly, whether the wings were made of paper or foil makes almost no difference to heat gain, even though these substances differ in their ability to conduct heat. What matters most is overall body size. Although all winged models absorb heat regardless of their absolute size, the impact is greatest on small models. In fact, bigger models gain less heat than smaller models.

Why do any of the models gain heat? For models with wings made of construction paper, thermal conductivity is low; this parallels the situation of real wings in modern insects. Adding wings increases the surface area of the insect, and the enlarged surface area captures heat more rapidly than the model is able to lose by convection, for a net gain in temperature. This makes sense with paper wings, but I wondered why it worked with aluminum foil wings, too. Koehl explained. Foil wings increase the surface area and, because metals are highly conductive, improve the wing's ability to both gain *and* lose heat. The net change in heat is thus about the same as in the experiment with paper wings. What advantage would having a higher body temperature bring? In insects, as in cold-blooded reptiles, raising body temperature by absorbing the sun's heat allows the insects either to be active longer or to engage in more vigorous types of activities without increasing the demands for metabolic energy. The advantage is so real that modern insects living in areas with fluctuating temperatures still have adaptations for elevating their body temperature passively. These experiments show that both evolving wings in the first place and enlarging those wings, once they existed, would offer significant improvements in thermoregulation. "Another paper suggested that we said that proto-wings are solar collectors," says Koehl with a grin, "but we never said that." However, their work shows that it was physically possible for proto-wings to have affected body temperature, not unlike a solar panel.

What about the aerodynamic properties of insect wings of various sizes and shapes? Using the same models, Koehl and Kingsolver measured the drag produced in a wind tunnel when the models were placed at right angles to the direction of the airflow. This arrangement approximates the situation of a horizontally oriented insect in free fall, and an increase in drag would mean a beneficial decrease in the speed of falling. They found that short wings (less than 20 percent of body length) have no significant effect on drag, regardless of body shape or the presence or absence of legs on the model. However, once wings are longer than 20 percent of body length, the amount of drag starts to increase steadily as wing length increases further. If the initial appearance of small wings does not affect drag and slow the velocity of falling, then the initial function of wings can't have been concerned with aerial performance. Yet once wings of moderate size had evolved, the evolution of progressively larger wings would impart aerodynamic advantages. This conclusion implies that insect wings were not initially parachutes but were reshaped into parachutes later on in the course of insect evolution.

Not all aerodynamic devices serve as parachutes, however, so Koehl and Kingsolver also looked at the function of wings in gliding. Could small wings prolong the horizontal distance traveled by an insect during a given loss in height? A more technical way of restating this question is to ask whether wings shift the glide angle (or the angle of descent) toward its optimum. According to the experiments by Kingsolver and Koehl, the answer for models with small bodies is "No." Wings on small-bodied models have to reach 60–70 percent of the length of the body before they improve gliding performance. The same is not true of larger models. For models more than six centimeters long, the answer might be "Yes," for wings as small as 30 percent of the length of the body improve the horizontal distance traveled, and for models ten centimeters long, wings only 10 percent of body length make a difference. Some researchers were so surprised by the striking differences in the performance of small- and large-bodied models that they questioned Koehl and Kingsolver's results, but their evidence has been supported by additional experiments and their calculations upheld. Koehl and Kingsolver agree that the models they used and the experiments they conducted were relatively simple, and using a more complex model—such as one with cambered wings—might change the effect of small wings, especially on small-bodied insects. But simple experiments with simple models are sufficient to detect major and powerful truths, and the results here are convincingly strong. In fact, the apparent paradox here (a wing in one model doesn't do what a wing in another does) is actually an example of a familiar and fundamental axiom in biology: the size of an animal has an enormous effect on its possible shape and adaptations. Although there is no *Archaeopteryx* of insects to indicate precisely how large the first winged insect was, probably no insect within the range of body sizes they investigated could gain an aerodynamic advantage in gliding or parachuting simply by evolving small wings.

The broad conclusions from Koehl and Kingsolver's studies are several. The initial appearance of small wings makes a big difference in the thermoregulatory abilities of insects of any reasonable size or shape. As wings increase in relative length, large-bodied insects experience aerodynamic advantages well before small-bodied insects do. Thus, if an insect simply increases its size, wings that were once too small to be of any use aerodynamically will suddenly become advantageous for flight.

Although their thermal experiments dealt only with wings held in one position, Koehl and Kingsolver explain that mobile wings are theoretically even more effective than fixed ones. Mobile wings can be positioned to fine-tune thermoregulation and, of course, are essential for the evolution of flapping flight. Thus they suggest that mobile wings may have evolved as thermoregulatory devices that were coincidentally preadapted for the later evolution of flapping flight. They conclude:

> [T]his scenario suggests that, during the period in which flight evolved, insects already possessed some thermoregulatory capacity. Thermoregulation and flight are intimately associated in present-day insects: a high and narrow range of body temperatures is necessary for vigorous flight in many insects. . . . We suggest that this association began at the origins of flight.

However, the pair is explicit about the limits of the kind of studies they have conducted. As Koehl wrote in a later article,

> [E]ven if such biomechanical studies show that a fossil function could have carried out some task or improved the performance of some function, that does not reveal the role that morphological feature served in the life of the organism; the best we can hope to accomplish with such quantitative studies is to reject functional hypotheses that are physically impossible.

In the end, demonstrating the plausibility or possibility of any particular hypothesis does not prove it is correct; the fossil record itself must be the final arbiter.

Do the principles of insect flight discovered in this research apply to the evolution of avian wings and flight? Maybe. Remembering some of the surprising results their experiments yielded, Koehl and Kingsolver caution, "One cannot predict a priori the functional consequences of small changes in a character." Conducting good experiments with bird models would be complicated by the presence of feathers that can be moved to alter the heat absorption and heat loss properties of the wing greatly, as Regal demonstrated. But it is worth remembering that wings can serve as thermoregulatory devices and, within a critical range, larger wings work better. Koehl and Kingsolver have also provided an interesting insight into a mechanism that may lie behind many evolutionary innovations: a simple shift in body size may change the function of an organ. Thus, if we knew how big "Proavis" was—and whether it was bigger or smaller than *Archaeopteryx*—we could try to speculate about the potential advantages of enlarging forelimbs into wings in the avian lineage. Unfortunately, "Proavis" is unknown in all respects, not just a few.

Although the debates about the initial function of wings and feathers can't be settled by considering this research on insect wings, some points are clear. Without question, a feathered proto-wing would be a most handy thing to evolve for thermoregulatory purposes. On the other hand, the specific aerodynamic impact of proto-wings on proto-birds is less certain. The earliest bird was almost certainly larger than the largest insect considered by Koehl and Kingsolver and probably had a rather different body shape, too. What may apply is the principle that wings must reach a minimum size (relative to body size) before any significant improvement in aerodynamic performance occurs.

Were the wings of *Archaeopteryx* long enough to function aerodynamically? I couldn't find an answer in the paleontological literature, so I decided to plunge in and find out for myself. A pertinent issue was the length of the proto-wings or forelimbs of "Proavis" and the other ancestors of *Archaeopteryx*. "Proavis," being a hypothetical animal, couldn't be measured, but I could bound my ignorance by examining the relative forelimb lengths of species that might represent animals of the same shape as the ancestors of *Archaeopteryx*. Because any potential ancestor of *Archaeopteryx* had a shape rather different from the insect models used by Koehl and Kingsolver, I could not measure wing length and body length exactly as Koehl and Kingsolver had. Using careful, scaled drawings by artist/paleontologist Greg Paul, I measured wing length as the length of an outstretched fore-limb (determined by adding the lengths of the forelimb bones—the humerus, ulna, and wrist bones—to the length of the longest finger, includ-ing the claw); I measured body length from the nose, along the skull, and down the curvature of the spine to the tail tip. With this technique, I found that *Compsognathus* and *Deinonychus*—two theropods commonly used as models of the body shape of the ancestor of "Proavis"—had forelimbs that were about 28 percent of their total body length. Insect wings start to func-tion in both an aerodynamic and a thermoregulatory role when they are over 20 percent of body length. Thus, if the same guideline applies to birds and their ancestors, the theropod forelimb from which the proto-wing of "Proavis" started to evolve was already long enough to offer advantages in both thermoregulation and aerodynamics. In fact, I realized quickly, this situation would pertain to almost any quadrupedal animal with the excep-tion of a few who have notoriously small forelimbs, such as *Tyrannosaurus rex*. Of course, the simple, unspecialized forelimb of the average quadruped doesn't offer much of an aerodynamic advantage—not enough, for example, to keep children from injuring themselves when they leap out of trees, pretending to be Superman. But the longer the proto-wing gets past 20 percent of body length, the better its aerodynamic and thermoregu-latory functions will operate.

By the time *Archaeopteryx* evolved, the forelimb had lengthened consid-erably over theropod proportions. In the Berlin specimen, the forelimb is about twenty centimeters long, slightly more than 50 percent the length of the nose-to-tail measurement (38.8 centimeters). Of course, even the parti-sans of the dinosaurian ancestry of birds do not claim that either *Compsog-nathus* or *Deinonychus* evolved directly into *Archaeopteryx*; I am using them as stand-ins for the unknown theropod ancestor. If there were a suitable crocodylomorph, the same comparisons could be made in the framework of that hypothesized ancestry. However, if the theropod-to-*Archaeopteryx* tra-jectory even roughly parallels what actually happened during the course of evolution, then the wing in *Archaeopteryx* offers a major increase in both aerodynamic and thermoregulatory effects over the forelimb of its ancestors.

But is length the critical issue? In Koehl and Kingsolver's experiments, length was a simple variable to change, while wing breadth was constant. In the case of *Archaeopteryx,* the stunning evolutionary difference is not that its forelimbs are so long but that they are so broad, enlarged by their feathers. More important than length are combined parameters like wing surface area (length multiplied by breadth) or aspect ratio (total wingspan, including the intervening body strip, divided by wing breadth). These ratios tell more about either heat transfer and airfoil function than length per se. Unfortunately, these parameters are difficult to calculate accurately in almost all fossils, except *Archaeopteryx*, because wing surface area and aspect ratio are measurements of the wing, not of the limb. While it is easy to put fossil bones together and add up their total lengths, only specimens with extraordinary preservation (like *Archaeopteryx*) show the dimensions of the wing itself. Indeed, the issue in the early stages of avian evolution was not simply lengthening the skeleton of the forelimb or proto-wing but broadening them, by adding those wonderfully light, insulating, mobile, and strong structures that are feathers.

Another biologist captivated by insects, Jim Marden from Pennsylvania State University, offered me a slightly different perspective on the functions of wings. Marden's insight arose by accident, during an early winter field trip to collect insects with his students. He is one of those field biologists who just can't help observing any species—large or small—that ranges within eyesight; life in all its glory is fascinating to him. On this particular trip, his attention was drawn by a new hatch of stoneflies in a mountainside stream that were, he thought, struggling very hard to fly. Stoneflies are common in aquatic locales, so Marden wasn't surprised to find them. They look somewhat like their ancestors in the late Carboniferous some 330 million years ago, except that the modern species has larger wings; they are still terrible fliers. On this occasion, Marden looked closer and realized that they weren't even trying to fly. They were doing something bizarre: standing on the meniscus at the top of the water and flapping their wings vigorously. This frenzied activity wasn't enough to produce liftoff. Instead, they buzzed rapidly along the surface of the water—they were surface-skimming. No one had ever realized the importance of this behavior before. By resting their feet on the meniscus, the stoneflies no longer had to support their entire body weight with their wings. They were taking advantage of the interface of two media, air and water, with radically different densities. The denser water supported the stoneflies; the less dense air made it possible for them to propel themselves by flapping rapidly, thus skimming across the water's surface to the stream's edge where they could crawl out to feed. All of the wing movements mimicked those used in flapping flight, but no true flight was involved.

Marden believes the stonefly probably *can't* flap its wings fast enough to take off and support the body's weight during cold weather—or, if they can, they don't. Speaking for himself and student Melissa Kramer, who soon became involved in the project, Marden says, "We have observed thousands of individuals in the field, but we have never seen one fly," except during warm spells. Why? It was too cold. The ambient temperature in the early winter along the streams of central Pennsylvania ranges between 0° and 12°C; because stoneflies are cold-blooded, this was their approximate temperature, too. They provide a direct example of the advantages of warm-bloodedness and thermoregulation. In the laboratory, where temperatures were much warmer (about 22°C), Marden and Kramer observed a different outcome. Out of thirty-one individual stoneflies that attempted takeoff in the lab, only six individuals (19 percent) were able to gain altitude. Another nine stoneflies (29 percent) could flap effectively enough to sustain level flight if they were released into the air, making a total of 15 (48 percent) who were able to fly in some fashion. But the largest proportion of their sample—the remaining sixteen (52 percent) stoneflies—couldn't fly at all. When they were released into the air, the stoneflies simply lost altitude until they encountered the water, a table, or the floor. Marden had previously studied aerial performance in birds and knew that the size of the flight muscles relative to total body weight was a key predictor of flying ability. He immediately wondered if the same was true in stoneflies. As he thought, the incompetent fliers had smaller flight muscles (relative to total body weight) than individuals who were able to take off or at least maintain altitude once launched.

Though flight was difficult for even the best fliers among the stoneflies, surface-skimming was much easier. Every stonefly—even those with the smallest flight muscles—could skim along the surface of the meniscus. In fact, stoneflies were able to surface-skim even in the cold and even if their wings were clipped to remove 70–80 percent of the original surface area. Because surface-skimming is clearly much easier and much less costly in energy than flying, Marden and Kramer theorized that surface-skimming is an important strategy—a locomotor mode that bridges the transition between swimming and flying. Seen in evolutionary terms, surface-skimming is easier than flying but requires exactly the anatomical equipment and the same neurological and muscular activity patterns. In short, surface-skimming preadapts its practitioners for flapping flight.

Marden and Kramer's results suggest an evolutionary scenario that starts with an aquatic insect with gills. To feed or mate, the insect needs to move around, so it paddles with its gill plates. Larger gill plates produce more movement, leading to better nutrition and more offspring who inherit the tendency for enlarged gill plates. In time, these gill plates enlarge to become proto-wings, which are in turn even more effective at underwater swimming. As the proto-wings increase in size further, to improve swimming capability, they eventually reach the threshold at which surface-skimming becomes possible. Surface-skimming is a marked improvement in the efficiency of movement. For the same expenditure of energy, a skimming insect can move much farther and faster than a swimming one. To skim more efficiently, the insect lineage evolves bigger wings and improves the nervous and muscular control over those wings. Finally, a species evolves that has all of the necessary anatomical equipment and locomotor behavior for full flapping flight. While flying is not more efficient or cheaper than skimming, it enables insects to take advantage of aerial and terrestrial resources that were previously inaccessible. Flying opens up a whole new world of possibilities, with many novel ecological niches.

As Marden and Kramer see it, this hypothesized evolutionary transition from swimming to skimming to flying parallels the developmental transition

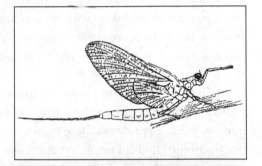

that occurs within the life of many modern insects, who have wingless, aquatic nymphs that metamorphose into winged adults. One more point supports this idea: stoneflies have hairy wings. The hairs trap air and help keep the wings dry, a useful feature for an insect skimming close to the surface of the water; wet wings are heavier and harder to flap. But, for an insect that flies all the time, hairy wings would be literally a drag. The hairs disrupt the smooth flow of air around the airfoil, creating turbulence that lessens the efficiency of the wing. Thus, in the mayfly, a near-relative of the stonefly but a better flier, the presence or absence of hairs on the wings is related to the de-

Figure 45. Mayflies similar to those studied by James Marden and Melissa Kramer (top) are also preserved in the Solnhofen limestones (bottom).

171

velopmental stage of the individual (Figure 45). Mayflies have hairy wings only when they are young; at this age, mayflies stay in contact with water and are weak fliers. As they reach maturity, they lose the hairs from their wings and are transformed into stronger-flying adults.

The intersection of the results of the work by Koehl and Kingsolver with that by Marden and Kramer demarcates a fascinating set of observations. Both studies emphasize how differently the same organ may function at different times during the course of evolution: a gill plate one day may be a wing/propeller the next; a solar panel may enlarge to be transformed into a useful airfoil. Wings, like feathers, may be thermoregulatory or aerodynamic in function, or both. This should come as no surprise, because feathers in the aggregate are a wing—and no one has supposed that any animal evolved only a single feather. The factors involved here are respiration, thermoregulation, and locomotion; they have enormous effects upon an animal's energy needs and upon the ecological niches it can exploit. At the most basic level, every organism must breathe, control its temperature, and move during at least some portion of its life. These are some of the non-negotiable demands of the organism, the essentials for life. Yet these crucial functions are all mixed up with each other. Changing one involves changes in all of them, because these functions are performed by the same structures, sometimes simultaneously, in a deep evolutionary and physiological tangle.

At first the great complexity of the shifting and multiple functions of these structures made me despair of ever straightening out the truth of the matter. But as I pondered these insect studies and their implications for the origins of bird flight, I came to appreciate a more fundamentally important point. The truth is that the functions of biological organisms *are* complex and are always intertwined. It is foolish to expect otherwise, for an organism must be functionally integrated and must remain so even during the process of evolutionary change, if it is to survive and prosper. This principle applies equally to insects and birds and their ancestors as each evolved flight in its separate way. The key is not to retain the same balance among these functions but simply to maintain a viable one. Modern birds present us with a complete package, an integrated system of respiration, locomotion, and thermoregulation that *works*. But, in all probability, this package arose piecemeal, as an ever-changing mosaic that did not take on its present configuration at the outset. The issue is not to imagine (or document) a single evolutionary step that transformed "Proavis" into *Archaeopteryx*, followed by another single step that transformed *Archaeopteryx* into a modern bird. The challenge is to document a process, a long series of changes and shifts in structure and function that only in retrospect form a trajectory from "Proavis" to birds. To trace such a complex sequence of events, I needed to identify the starting point, the anatomical and behavioral template from which *Archaeopteryx* and all later birds began. Within what group of reptiles would I find the ancestor of birds?

Chapter 8.
One Fell Swoop

Looking back through time from modern birds to the fossils, *Archaeopteryx* looks like the first bird—just an early flap in the evolutionary journey that produced the ultimate flying animal. But this teleological perspective suggests an intentionality or direction to the evolutionary process that was not present. Birds are exquisitely adapted for a dual locomotor system (hindlimb bipedalism and forelimb flapping), featuring specially adapted feathers and a turbo-charged respiratory system complete with air sacs. However, it is a serious mistake to envision *Archaeopteryx* as a proto-bird or an almost-bird or a future-bird. *Archaeopteryx* was a fully adapted *Archaeopteryx*, whatever type of animal that was. *Archaeopteryx* may have been in the trees, clambering awkwardly with its manual claws while striving to protect its feathers from damage, or running along the ground, leaping after prey with its feathered arms outstretched. I hoped knowing *where* it lived and came home to roost would help me discover *how* it lived and from whom it evolved.

The argument about the origin of bird flight and *Archaeopteryx*'s habitat is often characterized as "ground up" or "trees down," as if the only issue were the location of *Archaeopteryx* (or "Proavis") when it performed the activities that we see as beginning to fly. But "beginning to fly" is not a useful adaptation to any creature; whatever *Archaeopteryx* was doing, and wherever it was doing it, *Archaeopteryx* must have had a locomotor pattern that worked as it was, not one that would eventually evolve to be useful and practical.

Thus, as I evaluated these two competing theories, one criterion I used is whether or not each proposed stage in the evolution of bird flight could be shown to be a successful adaptation in its own right. Another issue is whether the transition between proposed evolutionary stages seems feasible and relatively easily undergone, or whether moving from one stage to the next requires a wholesale reorganization of the anatomy and movements of avian ancestors in a single step. The ease of transitions can be evaluated in part by considering the biomechanical constraints of various types of airborne locomotion, and what the requirements (in terms of anatomical equipment) are of each. Finally, I looked for internal flaws in logic in each theory. Sometimes these are assumptions difficult to reconcile with known facts; at other times, they are weak points where the explanatory power of the theory seems to falter. Each theory consists of not simply main tenets but a bundle of intertwined ideas and corollaries that round out the hypothesis.

Inevitably, the two theories contrast sharply on some points and are congruent on others and the details of each vary somewhat according to which scientist is speaking. But, as I discovered, this is treacherous ground for the outsider. Because the theories are two in number, the debate is structured as a choice between the two sides of a dichotomy; if you are "against" one theory, then you are "for" the other. Thus participants in the debate have often chosen their position not by endorsing that theory which they believe to be true, but by allying themselves against that which they believe to be false. In this case, the key factor in determining a particular scientist's stand on the origin of bird flight is often the extent of the weaknesses in the rejected theory, not the strength of the supporting evidence in the endorsed one. From a philosophical standpoint, this is an odd and sometimes unsatisfying way to proceed. It is also a particularly dangerous choice if there should turn out to be a third, as-yet-unimagined alternative to these two theories.

The "trees down" hypothesis has a major strength: it explains how takeoff could be accomplished by a species only just beginning to develop aerial locomotion. Since takeoff is the first difficulty to be overcome in any flight, explaining how *Archaeopteryx* or "Proavis" evolved the ability to fly is much easier to imagine if it could take advantage of a raised perch and the force of gravity. In contrast, the ability to take off from the ground is much more difficult, and explaining its evolution poses a major problem for proponents of the "ground up" hypothesis. Thus the "trees down" hypothesis makes a strength out of the greatest weakness of the cursorial hypothesis, reinforcing the structure of the debate as a choice between two branches of a dichotomy.

While living in trees neatly solves the problem of takeoff for *Archaeopteryx*'s ancestors, arboreality raises the opposite problem: unintentional takeoff or plummeting. I wondered whether this is a real or only a hypothetical problem. Do tree dwellers actually fall out of the trees enough to pose a significant danger? Searching the literature on animal locomotion, the best data I could find on the matter concerned gibbons. Gibbons, the small, long-armed apes of Southeast Asia, are some of the most adept and skillful arborealists among the primates. Nonetheless, in a survey of 118 gibbons shot in the wild, the primatologist and anatomist Adolf Schultz found that 36 percent of their skeletons showed one or more fractures that had occurred during life. Since an unknown, additional number of animals must have received injuries from falling that caused their immediate or slow death, the overall rate of plummeting is obviously substantial, and the potential danger of falling from the trees real. If gibbons fall out of the trees so often and with such devastating consequences—for a broken arm or leg is a serious handicap to a wild animal—the risk for a less skillful arborealist must be formidable. On the other hand, the negative consequences of unplanned downward trajectories would provide a potent form of natural selection, strongly favoring the survival of those who evolved or mastered some means of controlling the fall. For arboreal animals, becoming airborne occurs in one fell swoop; the rest of the process is finding or evolving postures and anatomical adaptations to control the swoop.

The idea of an arboreal origin for birds first came into favor in the 1920s, following the publication of *The Origin of Birds,* by Gerhard Heilmann. This synthesis was so masterful that it was considered the last word on the subject for many decades. The simple fact that birds are often found in the trees renders the idea attractive. It is wonderfully plausible that trees are the ancient and original avian habitat, given what birds do today. Positing an arboreal lifestyle for the ancestors of birds, or for the first birds, not only solves the problem of takeoff neatly, it also explains why aerial locomotion would offer immediate advantages. Aside from the obvious incentives for controlling falls, arboreal animals must move from tree to tree, for sooner or later any particular tree temporarily runs out of food, whether the animal in question is eating leaves, fruit, buds, gum, twigs, or insects. The more efficiently a species can move between trees, the better it will fare in evolutionary terms. More efficient locomotion may mean, for example, having the pick of the best food or the best nesting spots; another possible advantage is obtaining more food for the same expenditure of energy; and a third potential advantage is simply having more energy (or more time) left over for other activities.

Aside from outright flapping flight, the options for tree-to-tree travel are several. Where trees are closely spaced, there is often a sort of elevated highway of branches that can be followed in relative safety. Good balance, small size (so that small branches can be safely utilized) and grasping hands, feet, and, if possible, a prehensile tail make this sort of branch-to-branch locomotion more efficient. But such dense stands of trees are not common in all habitats and, even where they are, tree dwellers may need to travel considerable distances to find another tree in fruit or flower. Leaping from tree to tree is another possible mode of travel, if the distance between trees is not too great. Leaping is often much quicker than clambering along branches, where conditions are suitable. Vertical clinging and leaping is a distinct locomotor mode that is typical of many arboreal species; it was first described among the prosimian primates, a group that includes the tarsier, the small Asian primate with enormous eyes and big ears, the African bushbabies or galagos, as well as the Madagascan lemurs. Vertical clinger-and-leapers use both their hands and their feet to grasp a vertical support, with their bodies erect. Some vertical clinger-and-leapers have tails that are used like rudders, to help the animals change direction in mid-leap, while others seem to function perfectly well without the assistance of tail-rudders. Vertical clinger-and-leapers have large, muscular legs, and the bony joints of their knees and ankles are adapted both to enhance the mechanical advantage of the muscles that power the leaps and to restrict movement at those joints to a fore-and-aft plane. When moving between trees, a vertical clinger-and-leaper springs from one trunk, revolves around a vertical axis in the air, and lands facing the next trunk, feet first. During landing, the knees and ankles can flex and absorb the shock of impact before the hands reach out to grab the new vertical support (Figure 46).

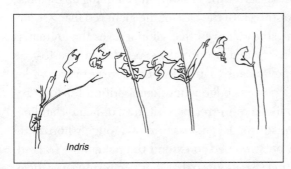

Indris

Figure 46. Vertical clinging and leaping is an effective means of moving from tree to tree that is typical of some living prosimians, like this *Indris* from Madagascar.

Another effective means of traveling from tree to tree is gliding, if an animal has the appropriate anatomical adaptations. Gliders are invariably small—less than about 1.5 kilograms (three pounds) and sometimes much smaller—and have a body that is flattened back to stomach. Gliders also must have some form of airfoil and often extend some part of their anatomy sideways while gliding, to increase the surface area of the airfoil. For example, some gliders have a sim-

ple flap of skin, like the erectile neck frill and ribs that can be stretched sideways of the flying lizard *Draco* (Figure 47). Another option is to have a full-fledged patagium that stretches from arms to legs, as found in flying squirrels and many other gliding mammals, while flying fishes use their enlarged pectoral fins as airfoils. True gliders cannot flap, but many flapping fliers glide—or, more properly, hold their wings outspread without flapping—temporarily. Compared to climbing down

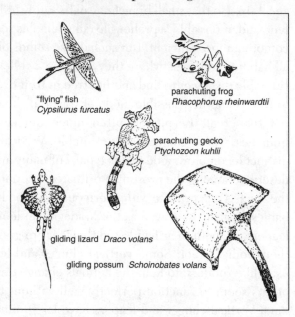

Figure 47. Gliding animals are typically small and have patagia or some other type of airfoil to enlarge the surface area of their bodies and let them prolong their glides.

from one tree, walking or running across the ground, and climbing up the next, gliding is a wonderful mode of travel. In the case of the giant red flying squirrel, *Petaurista petaurista,* one study showed that gliding saves the animal as much as 25 percent of the energy that is normally expended in tree-to-tree travel. This energetic efficiency means that gliders can forage over a wider area or can feed faster, an important edge where competition for food resources is fierce. However, gliding tree-to-tree poses its own difficulty: landing. After takeoff, an archetypal quadrupedal glider assumes a roughly horizontal position, with arms and legs outstretched to extend the patagium. To land, such gliders must perform a tricky stall-out so that they can resume a vertical, head-up/tail-down position suitable for clinging to the vertical support.

Unfortunately, few habitats offer trees so closely spaced that gliding is the only mode of tree-to-tree travel needed throughout an animal's entire lifetime. Most gliding animals must face a perilous transit across the ground at some time, followed by a climb up a new tree trunk to regain gliding height. The terrestrial portion of the endeavor is dangerous because the glider may be exposed to swift-moving ground predators. This is a real problem to a small gliding animal, for the terrestrial locomotion of patagium gliders is hindered by the patagium itself. The patagium prevents them from moving fore- and hindlimbs in a rapid diagonal gait like the trot. The patagium also seems to inhibit galloping or rapid hopping in which both forelimbs move together, followed by both hindlimbs. Instead, patagium gliders often move in a gait like that of an inchworm or a slow-moving kangaroo, although every second spent on the ground prolongs their vulnerability to predators. Once the next tree is gained, the glider must begin the climb up the trunk again. In sum, gliding involves a rapid initial movement in which a considerable horizontal distance is traveled while vertical distance (height) is lost, followed by a slower phase in which vertical height is regained while very little horizontal distance is traversed. As Dieter Stefan Pieters of the Forschungsinstitut Senckenberg in Frankfurt observes: "Climbing certainly has to be regarded as an essential ability of an arboreal animal. Even a glider has to return somehow to its arboreal environment. In this respect it is hardly imaginable how flight-feathers should evolve on a forelimb used for climbing." Even for species considered to be gliders or vertical clinger-and-leapers, both climbing and some form of terrestrial locomotion are essential parts of their adaptations. Although both of these locomotor modes harness the free power of gravity for takeoff, some energy must be expended in the initial leap or dive out of the tree and still more must be used to steer and position the animal for landing. Gliding is not a free ride.

The word "gliding" is itself somewhat misleading, at least for English speakers. Animals with patagia glide in the proper sense—that is, move from one raised support downward to another substrate; they use anatomical and postural mechanisms to increase their effective surface area, which slows the downward trajectory and can be used to help steer a desired course. The airfoil helps prolong and slow the glide, but it does not power the movement. Gliding also necessitates some means of climbing and terrestrial locomotion. The contrast between gliding and flapping flight is obvious, for only the latter requires a flight or power stroke.

However, animals like birds or bats that engage in active, flapping flight may also do something that is often called gliding but is really a temporary suspension of flapping. The "gliding" practiced by flapping fliers is fundamentally different from true gliding, although unfortunately the same word is often used for both in English. More accurately, this activity ought to be called soaring or aerial coasting, by analogy to a bicyclist who temporarily stops pedaling, rather than gliding. Flapping fliers can take advantage of the air movements of natural updrafts and currents to provide propulsion, at least intermittently, which saves energy. The albatross is a specialist in this regard; its huge wingspan permits it to soar for long periods, without flapping, as it rides the air currents produced as the waves deflect the wind. This strategy is known as slope soaring and can also be used to ride the wind currents as they move up a hill or ridge over land. An idea of just how effective soaring over the waves can be in terms of saving energy can be gleaned from considering some of the remarkable measurements of single foraging trips by albatrosses. Using radio telemetry, two researchers tracked five male albatrosses who left their mates ashore incubating eggs while they flew off to forage for food. A single trip lasted anywhere from two to almost five weeks. Individual birds covered up to 554 miles a day for a total of between 2,257 and 9,363 miles per trip. The birds flew at speeds up to forty-nine miles per hour, mostly during the day, and rested on the water's surface for no more than an hour and a half at a time even at night. After these heroic feeding trips—which surely would not have been possible if the albatrosses were flapping continuously—the males returned to the nests to incubate the eggs, freeing their partners to go on their own foraging trips.

Vultures are not known to perform such extraordinary feats of flying, but they are classic slope-soaring birds who also exploit thermals, the naturally occurring spirals of warm air that rise over land. A vulture riding a thermal moves in circles, to stay within the column of rising warm air, and can survey large areas for carcasses without expending much energy. Smaller birds also practice energy-saving strategies that involve the intermittent cessation of flapping. For example, gulls, crows, and other medium-sized birds fold their wings every few meters during flight, producing an undulating flight path. Still smaller birds like tits or wagtails may coast every few wingbeats, flapping diagonally upward on a straight path and then coasting downward with their wings outspread but unmoving, producing a pattern called bounding flight. Whatever the bird's size, it cannot migrate without using some means of resting in the air and conserving energy. But intermittent nonflapping flight, as practiced by flapping fliers, is a distinctly different strategy from the obligatory gliding practiced by flying squirrels, sugar gliders, or the colugo. Technically, gliding involves a gravity-powered, nonflapping, downward descent at an angle of less than forty-five degrees to the horizontal. Thus the anatomical and behavioral adaptations required by flappers-who-sometimes-soar are different from those of quadrupedal animals who sometimes glide. Failure to distinguish clearly between true gliding and intermittent flapping strategies has caused much confusion in the attempts to understand the origins of bird flight.

This distinction is especially important in considering the arboreal hypothesis, because an essential element of that hypothesis—perhaps its main component—is that gliding was a precursor to flapping flight. Following a seminal paper by Walter Bock of Columbia University in 1965, the proponents of this hypothesis envision an evolutionary trend toward an increasing mastery of aerial movements, each step of which is a distinct and viable adaptation. As a first step, avian ancestors were arboreal animals who occasionally engaged in accidental free fall (plummeting) and evolved the ability to parachute. Parachuting is defined as involving a downward trajectory of more than forty-five degrees to the horizontal; parachuters typically use some mechanism to increase surface area, thus slowing the animal's velocity. Parachuting does not have to involve anatomical changes, for postural adaptations like spreading the arms and legs will deploy whatever patagium or extra flesh lies between each arm and leg. Another postural component to parachuting is erecting the fur or feathers, which increases the surface area of the airfoil. Simple postural adaptations for parachuting can be surprisingly effective. Although house cats are not generally regarded as masters of aerial locomotion, a study of cats brought to a veterinary clinic in New York City after falling out of windows revealed that cats have some important parachuting behaviors. Statistically, cats who had fallen from three or more stories survived at a better rate than those who fell shorter distances. The explanation for this apparent paradox was that cats who fell from lower windows did not have enough time to rotate in the air and adopt a parachuting posture before they hit the ground. In contrast, cats who fell from greater heights reduced their terminal velocity by spreading their limbs and erecting their fur, and they received fewer deadly injuries.

Animals that have explicit adaptations for controlling descent do even better. Folds of skin between the fore- and hindlimbs that are exaggerated into patagia impart an impressively greater ability to slow a fall. Leaps—controlled falls—can be undertaken as a matter of choice once such adaptations are in place. In the "trees down" theory, parachuting is thus seen as providing a smooth transition into deliberate gliding, in which the downward course and speed are lessened, the horizontal distance covered is prolonged, and steering becomes better controlled. Several factors determine whether an animal parachutes or glides. The glide angle itself is established by the ratio of lift—the upward movement produced by the wings, airfoil, patagium, or whatever—to drag, the resistance to movement through air (or any other medium) that, combined with the action of gravity on the animal's weight, produces a downward movement. When the ratio of lift to drag is less than one, the animal follows a steeper path in the air and parachutes; where the lift/drag ratio is greater than one, the trajectory is shallower and the animal glides. The amount of lift in turn depends on the size, shape, and angle of attack of the airfoil, so all of these also affect whether parachuting or gliding is possible. So, too, does overall weight. Because terminal velocities of greater than six meters per second may be literally *terminal*, or at least injurious, animals that weigh more than about eight grams (about two ounces) simply cannot succeed as habitual parachuters.

Gliding is "intuitively attractive" as a precursor to flapping flight, at least in the mind of Jeremy Rayner, an English biologist who specializes in studying flying animals.

> Gliding allows gravity to help rather than hinder the development of flight, it assumes that flapping first developed at high speeds, and it assigns more importance to the relatively simple requirements of forward flight than to control, maneuverability or prey capture. . . . Gliders—without [wing] elevator muscles— already possess the musculature for flapping at normal flight speeds (although the muscles themselves may not be sufficiently large), and therefore are preadapted for the development of true flapping flight. Once level fast flight is achieved, pressures for greater control, for slower flight or hovering, for more advanced manoeuvres, and for improved landing would favour development of elevator muscles and greater upstroke control.

One great strength of the arboreal hypothesis derives from living animals who move through the air. This theory relies on a logical sequence of adaptations involving an increasing mastery of the air: plummeting, parachuting, gliding, and flapping. Each stage is a viable locomotor mode in and of itself, one that is practiced successfully by living species. There are no awkward "almost-flying" stages along the hypothesized evolutionary transition to raise the question of how a creature moving in this way could have survived for a week, much less long enough to leave successful progeny. It is a simple mental leap from recognizing this ordered sequence of living, but unrelated, species to postulating an evolutionary trajectory for a single lineage that moved through these stages over time, eventually producing a modern bird. Indeed, Bock used his paper in 1965 to articulate the advantages of each stage in this model of the evolution of birds. In his scheme, a quadrupedal ancestor first became bipedal and then terrestrially cursorial. The hypothetical "Proavis" was identified by Bock as the next form, which was both bipedal and arboreal. From there, the avian lineage progressed through leaping to parachuting and eventually to true gliding, the way of life most proponents of this theory envision for *Archaeopteryx*. Eventually, the avian lineage abandoned gliding and evolved the true flapping flight characteristic of modern birds. As Bock and many others who favor this theory observe, the beauty of the arboreal theory is that each stage works perfectly well, and the locomotor patterns become increasingly aerial by small increments. It is a neat and appealing scenario to some. For example, Alan Feduccia, reviewing the many flying vertebrates that are alive today, summarizes his view of the situation: "The one thing they all have in common is 'trees down' or 'high places down.' In order to evolve flight, you need small size and high places. You've got to be up to be able to take advantage of the cheap energy provided by gravity." To him, this argument is so compelling that "almost nothing" could disprove the "trees down" hypothesis.

Another generally cited strength of the arboreal hypothesis, as exemplified by Bock's presentation, is that at each of these postulated stages, takeoff is demonstrably possible. There is no problem getting into the air from an arboreal roost; the only problem is not dying when the ground or some other substrate is encountered at landing. There are strong echoes of many of the early experiments of human aviators who used barn roofs, cliffs, and towers as takeoff points—and often, tragically, the ground beneath them as death sites. The problem is that, once takeoff is solved, landing becomes an acute and serious issue. Although the very adaptations that transform accidental free fall into parachuting or gliding lessen the danger of abrupt landings, those landings may nonetheless be deadly because they involve drops from a substantial height. In contrast, taking off from the ground may be inherently tricky and may make attaining height difficult, but miscalculations have far

183

less serious consequences when the ground is close beneath you, as the Wright brothers discovered. In short, the difficulties of takeoff and landing are reciprocal. When takeoff is easy, landing is potentially dangerous; when takeoff is difficult, landing is easier, or at least less often fatal.

Unfortunately, the arboreal theory is not wholly convincing, however. Its first and most obvious problem lies in the lack of an ancestor. Arborealists are united in rejecting theropod dinosaurs as bird ancestors, in part because they are stymied when it comes to envisioning how a bipedal animal built like a theropod got up a tree. The problem is, if theropod dinosaurs are rejected as potential ancestors, who is left? There is no convincing candidate among the known thecodonts, pseudosuchians, or crocodilians. Frankly, these are all poorly known and poorly defined groups of animals. Of course, the paucity of factual knowledge about these groups leaves the door wide open for speculation about what an as-yet-undiscovered species will look like; few possibilities can be eliminated. Unfortunately, the number of possibilities that can be convincingly defended is equally small. Arborealists are at present unable to summon up any plausible candidate for the ancestor of birds. Larry Martin describes his vision of the ancestor of birds by saying, "If you can imagine a small, bipedal crocodile that lives in the trees . . . ," but there is no fossil evidence that helps assemble these traits into a tangible creature. And although they may protest that their opponents are also unable to identify the particular theropod that was ancestral to birds, those who favor a dinosaurian ancestor can at least point to several different species from various times and places that show many of the features that would be appropriate for an avian ancestor. They have a general type of dinosaur to point to, if not an individual species, and those favoring the arboreal hypothesis do not. On this count, at least, defenders of the cursorial hypothesis have a greater weight of evidence on their side.

The next point of contention is Bock's evolutionary trajectory of increasing mastery of the air, which Kevin Padian for one finds seriously flawed methodologically. "Walter's method is to erect a pseudo-phylogeny," says Padian.

> Let's ask ourselves what a pseudo-phylogeny is. You assemble a group of animals that you think represent stages in the evolution of a trait [like flapping]. How could you test your pseudo-phylogeny? Are you going to assemble things that don't work? No. This method consists of simply actualizing the previous con-

ceptions of the investigator; it can do nothing else. So what does a pseudo-phylogeny tell us? Nothing.

He continues,

> The other problem arises because Walter doesn't think fossils are important to understanding how birds acquired flight or anything else. Now, if you're going to take that position and not look at fossils or other extinct animals, your possibilities are perforce limited by what we have alive today. But the present diversity of species is way less than the past diversity. How could you possibly rationalize ignoring a major part of the diversity? I don't see that pseudo-phylogenies are anything that comparative biology is going to accept as a method because it doesn't lend itself to testing very well.

Like Mimi Koehl, Padian believes that the ultimate test of any theory about evolution is the degree to which it matches the fossil record. After all, the fossil record is the only documentation of what actually happened during evolution.

Padian also objects because Bock's scheme ignores the distinctness of each of these adaptations. Gliding is simply not the same as soaring, a period of non-flapping within flapping flight. Soarers and flappers look the same and have the same anatomical adaptations, but this is because they are the very same species in many cases. Soaring is a secondary adaptation, a part-time locomotor mode practiced by species, usually of larger-than-average body size, who are primarily flappers. In contrast, a glider and a flapper look entirely different because they have very different adaptations for moving around the world.

Padian and Rayner have catalogued the anatomical differences between flappers and gliders. One major difference involves the length and structure of the forelimb. Though both gliders and flappers tend to have longer arms for their body length than do nonflying relatives, flappers hypertrophy and greatly elongate the outermost segment of their forelimb, while gliders never do. The anatomical differences reflect aerodynamic ones, for the outer wing segments provide the thrust or forward motion in flappers, whereas the inner two wing segments provide lift in gliders.

Gliders and flappers also differ dramatically in their characteristic wing and body shapes. Gliding is apparently not a difficult adaptation to evolve, for it has arisen at least seven times independently within mammals. Nonetheless, gliding seems to dictate a narrow range of habits, sizes, and shapes, for gliders show a remarkable convergence in body shape and anatomical design. Gliding animals are generally small, ranging in weight from about 0.33 ounce to 3.3 pounds (or 9 grams to 1.5 kilograms); they are also usually nocturnal, vegetarian, and predominantly arboreal species found in tropical forests. What does this mean in terms of their physical adaptations? Gliders have small wings consisting of patagia stretched between arms and legs, so that the outline of a typical mammalian glider, seen from below as it glides, is usually approximately square or rectangular (Figure 48). Because the wing is only an enlarged membrane, its total surface area is restricted and it produces a limited amount of lift. A standard measure of this property is wing loading, the ratio of the animal's body weight to the surface area of its wing. Wing loading is a quantitative way of answering the question, How much wing does this animal have to support its weight in the air? Gliding animals typically have high wing loadings, on the order of 1.2 grams per square centimeter. Another key aerodynamic parameter is aspect ratio, which is defined as the total wingspan (including the body strip that intervenes between the wings) divided by the wing breadth; animals with high aspect ratios thus have long, narrow wings, while animals with low aspect ratios have short, broad ones. Gliders

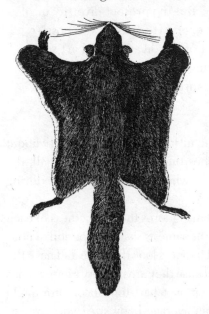

Figure 48. This copper engraving of a flying squirrel in the act of gliding shows the typical rectangular outline of a mammalian glider seen from below. Because the patagium connects the forelimbs to the hindlimbs, gliders move awkwardly on the ground.

have exceptionally low aspect ratios of 1–1.5. Other than those of body shape, gliding mammals show relatively few skeletal adaptations to gliding that are consistent across all groups. In fact, gliding mammals actually resemble flightless birds in having short wings, low aspect ratios, and high wing loads. For example, measurements of ten flightless birds revealed wing loadings ranging from 1.01–1.57 g/cm²—encompassing the range of mammalian gliders—and aspect ratios that were only slightly higher than those of gliders, ranging from

1.7 to 2. This resemblance shows how slight the specializations for gliding are, the most obvious anatomical feature being the possession of a patagium that is manipulated or controlled by movements of the fore- and hindlimbs. Judging from known gliders, two options seem to be theoretically impossible. First, no known four-limbed glider has ever evolved a forearm-only means of gliding. Four-legged walkers become four-legged gliders. Second, although it is theoretically possible for a glider to have a high aspect ratio—by evolving much longer arms than legs—no known species has such an adaptation.

In contrast, flapping flight is an anatomically demanding adaptation that has nonetheless evolved in animals of a wide range of body sizes and ecological habits. Flapping requires many distinctive skeletal specializations, such as changes in the shoulder and forearm anatomy, an enlarged and reinforced sternum, and the fusion of the clavicles into a furcula, to anchor flight muscles. Flappers have much larger wings (relative to body weight) than gliders, leading to low wing loadings and much higher aspect ratios. Most flying birds have wing loadings that fall within a range of 0.3–0.6 g/cm^2. Only a few exceptional species like golden eagles and albatrosses approach or match the high wing loadings of gliders, with values of 0.9 g/cm^2 and 1.2 g/cm^2 respectively. The lower wing loadings of flappers are coupled with higher aspect ratios, which in living, flighted birds may range from about 4.5 to 20.

The marked differences between gliders and flappers mean that evolving a flapper from a glider might involve almost as many anatomical changes as would evolving a flapper from a terrestrial biped or quadruped. The team of Russell Balda, Gerald Caple, and William Willis compared the aerodynamic and physical requirements of gliding and flapping and concluded that

> in any transition from a glider to a powered flier there is a non-adaptive transition zone. With primitive flaps of a gliding animal the glide path is not extended, in contrast, it is degraded. . . . A powered flier can easily become a glider, but a glider would face a most difficult transition to a powered flier. . . . an animal limited to gliding is unlikely to flap.

Their analysis suggested that a glider would have to abandon its ability to glide in order to evolve an ability to flap, unlike a soarer which can freely alternate flapping and "gliding." Their research thus suggests that there is an absolute barrier to the evolution of flapping flight from a gliding form. Their calculations would form a compelling argument but for one fact: bats have done it. Without dispute, bats evolved from a once-gliding ancestor. And, if bats could evolve flight from gliding, so could birds, at least in theory. Unfortunately, calculations and estimates concerning hypothetical evolutionary transitions do not necessarily indicate what actually occurred in the past; they only provide an educated guess at the likelihood of particular events having happened. While it may be unlikely that birds evolved from gliders, unlikely things have happened more than once in the course of evolution.

There are two practical ways to test the hypothesis that *Archaeopteryx* and its ancestors were tree-dwelling, and some ingenious studies have been performed. Most obviously, the arboreal hypothesis requires that there were trees to be climbed. Given the exquisite preservation of fossils in the Solnhofen limestone, the nature of the habitat and vegetation at Solnhofen in *Archaeopteryx's* time is a fairly simple matter to determine. Several scientists have made an intensive study of the Solnhofen limestones and the paleoenvironments they preserve.

All limestone—and the Solnhofen limestone is no exception—is made up of the fossilized remains of the calcareous skeletons of millions upon millions of tiny marine organisms. The Solnhofen limestones occur over a huge area about seventy by thirty kilometers, where the skeletons of marine organisms were deposited in shallow-water depressions between algal-sponge reefs. The general environment was a lagoon, with warm, shallow waters that were protected and usually still, although occasional storms dumped open ocean water and its inhabitants into the lagoon. Few life forms larger than bacteria and algae regularly lived in the lagoon, because the bottom layer of water was hypersaline and oxygen-depleted, unfriendly to life. Thus when organisms died in the lagoon or fell into it as carcasses, they soon sank into the stagnant bottom waters where they were undisturbed by scavengers. This accounts for the largely intact nature of many of the Solnhofen fossils, which are the ones for which postmortem disturbance was minimal. In addition to the wonderful Berlin *Archaeopteryx*, there is, for example, a fossil fish with a smaller fish halfway into its mouth. Everything, from scales to fins and bones, is in place. From the watery world, the Solnhofen limestones preserve sponges, jellyfish, annelid worms, mollusks of many types, exquisite ammonites, squids and cuttlefish, shrimps, crabs—including gorgeous horseshoe crabs—barnacles, starfish, sea urchins, sharks, rays, bony fish, chelonian turtles, ichthyosaurs, crocodiles, and plesiosaurs. From the terrestrial world that surrounded the lagoon come amazing fossils of insects and spiders, lizards, the small theropod *Compsognathus*—complete with its lizard prey in its gut—a fabulous array of pterosaurs from the tailed

Rhamphorhynchus to the tailless *Pterodactylus,* all with marvelous impressions of the skin wings. For once, this is a fossil assemblage close to the proverbial snapshot-in-time of an ecosystem, what paleontologists call a lagerstätt. Admittedly, the representation of aquatic and aerial species seems to be better than the preservation of strictly terrestrial species, which are sparsely represented. In fact, some of the terrestrial animals preserved at Solnhofen apparently drifted near the surface for days rather than sinking immediately. On the surface, they were not protected, so these carcasses decayed and disarticulated partially before sinking. These specimens were often swept by the prevailing wind currents into the western part of the Solnhofen lagoon before they sank and became fossils. Such specimens are the less complete fossils from Solnhofen, like the Maxberg *Archaeopteryx.*

Figure 49. The environment of the Solnhofen lagoon may have looked like this 150 million years ago, when *Archaeopteryx* was alive. There is no fossil evidence of tall trees in the area, despite the excellent preservation of plant fossils. There were shrubs in the area that may have grown as large as ten feet (three meters) tall.

Plants, too, are preserved in the Solnhofen limestone, and these are types that grow in arid habitats and salty soil (Figure 49). The most common land plants, *Brachyphyllum* and *Palaeocyparis,* were scaly conifers of the type known as stem succulents; only the central trunk was woody, while the branches and stems were fleshy, water-storing tissues. These plants cannot grow very tall and must have been shrubby or bushy in appearance, with many branches and a narrow main trunk. Günter Viohl of the Jura-Museum in Eichstätt, Germany, has summarized much of the evidence about the Solnhofen paleoenvironment and estimates their maximum height at about three meters (ten feet). In fact, there were probably no tall trees in the immediate vicinity of the lagoon, since no logs or sizable pieces of driftwood have ever been recovered from the Solnhofen limestone. The diameters of the thickest pieces of fossil wood from Solnhofen are smaller than about three to four centimeters (about 1.5 inches).

Viohl suggests that *Archaeopteryx* inhabited one or more of the small offshore islands dotted about the lagoon. These islands were covered in a mosaic of open plains and conifer bushlands. There may have been cliffs along the shore but the evidence is equivocal, and some scientists believe the islands were fringed by sandy beaches. Because the limestone has yielded fossils of insects like dragonflies, which need freshwater for their larvae, we know the islands must have had small ponds or at least puddles of freshwater at times. Freshwater was limited, however, and the islands lacked large permanent rivers that emptied into the lagoon. This point is shown by the scarcity of terrestrial sediments that would have been dumped into the Solnhofen basin if such rivers had existed. Without permanent sources of freshwater, these small islands cannot have supported large trees. Thus, if Viohl is correct that *Archaeopteryx* lived on the offshore islands, there were simply no trees—and not possibly any woodlands or forests—for *Archaeopteryx* or its immediate predecessors to live in.

On the face of it, the lack of trees is a fatal blow to the arboreal hypothesis because arboreal animals need arbors. However, the proponents of the arboreal hypothesis are not easily discouraged. For example, Feduccia points out that tree trunks or other large woody specimens are absent in many fossil sites, so their absence at Solnhofen "cannot be used to prove the absence of trees." Indeed, proving the *absence* of any species, plant or animal, at a fossil site is extremely difficult, given how inconsistently the living community becomes fossilized even in a lagerstätt. For example, other European sites that are roughly contemporaneous with Solnhofen have fossils of stegosaurs, brachiosaurs, sauropods, hypsolophodontids, and many other dinosaurs. Were these species absent from the Solnhofen region—perhaps because the environment was unsuitable for their needs—or present and unrepresented? There is no easy way to answer this question with certainty. Trees *may* have been present at Solnhofen. What works against Feduccia's argument, however, is the presence of fossils of plants and shrubs at Solnhofen. All other things being equal, larger, heavier, and more woody pieces are more likely to be preserved—if they were present—than smaller and more fragile ones are. Another possibility, offered by Feduccia and others, is that *Archaeopteryx* may not have lived in the immediate vicinity of the lagoon, the traditional interpretation. It is generally true that most fossil species are found dead where they lived, but *Archaeopteryx* might be an exception to the rule. Possibly *Archaeopteryx* fossils are individuals who were blown far off course and well out of their usual habitat by storms or monsoons. In support of this scenario is the fact that there are landmasses to the north of the lagoon that supported trees; in opposition to this scenario is the sheer improbability that seven remarkable fossils were preserved under such unusual circumstances, several of them without visible damage to their feathers.

Alternatively, perhaps an "arboreal" *Archaeopteryx* actually inhabited shrubs and bushes rather than trees per se. Even a modest elevation of a three-meter shrub offers the power of gravity for takeoff. If this speculation is correct, *Archaeopteryx* regularly took off from a low altitude and must have landed on the ground frequently. A few calculations will show how limiting a takeoff height of three meters, or about ten feet, is—assuming that *Archaeopteryx* started at the top of a shrub of maximum height for the Solnhofen area. In order to glide and not to parachute, an animal must achieve a glide angle of no more than forty-five degrees below the horizontal. If *Archaeopteryx* were a rudimentary glider with a glide angle of about forty-five degrees, *Archaeopteryx* would travel a horizontal distance exactly equal to the vertical distance descended: only about ten feet. With such a poor gliding ability, *Archaeopteryx* could avoid landing on the ground only if the shrubs were nearly continuous, in which case climbing along branches from shrub to shrub would work as well, although it would be slower. If *Archaeopteryx* were a good glider and could achieve a glide angle of thirty degrees, it would travel a maximum of only twenty feet horizontally. (Some gliders use a strategy known as "drop and flare," in which they first drop unimpeded to gain velocity and then spread and flatten their body to produce a low glide angle. While this is an effective strategy in animals like the red squirrel, a maximum starting height of ten feet precludes the use of flaring.) Thus it seems certain that an arboreal, gliding *Archaeopteryx* must have possessed good climbing skills and, if it lived in the Solnhofen area, could have glided for only very short distances.

Therefore, the second test of the arboreal hypothesis is to examine the anatomy of *Archaeopteryx* to see if evidence of climbing skills can be found. One focus is the forelimb of *Archaeopteryx*. Could its enigmatic wing claws be used to climb? Derik Yalden, a zoologist at the University of Manchester, investigated forelimb function in *Archaeopteryx* in response to work by John Ostrom. Ostrom originally observed that the claws of the hand of *Archaeopteryx* are strongly curved, compressed side-to-side, and needle-sharp, whereas its foot claws are straighter, less narrow, and less sharply pointed. From these features Ostrom deduced that *Archaeopteryx* was a terrestrial biped that used its sharply tipped fingers in predation. "As one who much prefers the concept of an arboreal ancestor for birds in general, and envisages *Archaeopteryx* as an arboreal animal," Yalden admits with disarming candor, "I found these arguments quite convincing, and therefore disturbing."

His unease prompted him to compare the claws of birds and mammals with those of *Archaeopteryx* to try to deduce the function of the latter. Although he did not quantify his observations, Yalden found very sharp, strongly compressed claws to be typical of trunk-climbing species, like woodpeckers among the birds or squirrels among mammals. Predatory birds, like the long-eared owl or the sparrow hawk, have strongly curved claws that are more conical in cross-section than those of *Archaeopteryx* (Figure 50). The exception

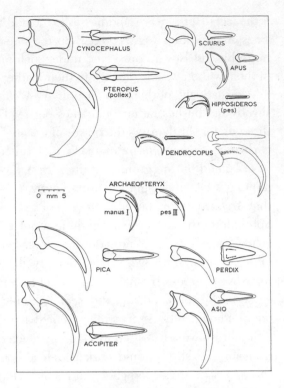

Figure 50. Derik Yalden compared the shape of *Archaeopteryx*'s bony claws and their bony coverings to those of various birds and mammals, concluding that *Archaeopteryx*'s sharp, narrow claws more closely resemble those of tree-climbing species than those of predatory ones. Trunk-climbing mammals are represented by *Sciurus* and *Cynocephalus*; *Pteropus* and *Hipposideros* are bats; *Apus* is a cliff-nesting swift; *Dendrocopus* is a tree-climbing woodpecker; *Pica* and *Perdix* are ground-dwelling birds; and *Accipiter* and *Asio* are avian predators that grasp their prey with their feet.

to the rule is that bats, which cling with their feet and may also grasp prey with them, have highly curved and strongly compressed claws. Because mammalian claws are constructed differently from those of birds, Yalden found the similarity of *Archaeopteryx*'s claws to bat claws less compelling than the resemblances to climbing birds. He concluded that *Archaeopteryx* probably was an inefficient flapper, launching itself from a tree and then climbing up the next one using its manual claws. Although many scientists were impressed with Yalden's observations, the Teyler *Archaeopteryx* has beautifully preserved claws—not just the underlying bone but the keratinous claws themselves are preserved. These show neither blunting nor wear, which would surely be expected if these claws were regularly used for climbing tree trunks (Figure 51).

Yalden also looked at forelimb proportions among various species, since gliding vertebrates (such as flying squirrels) generally have longer forelimbs relative to trunk length than do nongliding forms. In squirrels, the relative length of the hindlimb varies with body size; small flying squirrels have hindlimbs of normal proportions while larger flying squirrels have elongated hindlimbs. Yalden found that *Archaeopteryx* has an elongated forelimb and hindlimb relative to nongliding squirrels. The significance of this fact is unfortunately obscure, for it makes little sense to assess the relative forelimb length of an early bird by squirrel standards. Squirrels do not reveal much about nonsquirrels, since different groups of animals have different characteristic limb proportions—some are basically long- or short-legged, for example, while others tend to the forelimb-dominated or hindlimb-dominated, because of their particular evolutionary history. The only valid approach would be to compare the relative forelimb length of *Archaeopteryx* to forelimb lengths in a series of closely related but nongliding relatives—if only paleontologists could agree on exactly who those relatives were.

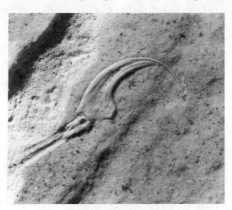

Figure 51. Arguing against Yalden's conclusion that *Archaeopteryx* used its claws for climbing tree trunks is the extreme sharpness of the horny claws of *Archaeopteryx*, seen here in a close-up of the Teyler specimen. Habitual tree-climbing ought to produce blunting and wear on the horny claw.

A few years after Yalden reported his work, Alan Feduccia undertook his quantitative study of the curvature of the claws of *Archaeopteryx* in another attempt to test the arboreal hypothesis and found that the foot claws of *Archaeopteryx* were closest to those of climbing or perching birds in terms of their curvature. But curvature is not the only foot adaptation that facilitates this type of behavior. Perching birds need both curved claws and a reflexed hallux (the opposable claw) to help lock their feet onto a branch. For their part, climbing birds like woodpeckers have very robustly built toes with strong muscle markings, features that attest to the sheer strength needed to climb (or walk) bipedally up a trunk. Such birds are also invariably zygodactylous, with two toes curving forward and two curving backward (Figure 52). This is an uncommon arrangement that is found in only a few groups of living birds, like woodpeckers (the family Picidae), puffbirds (Bucconidae), and jacamars (Galbulidae), as well as barbets, toucans, and honeyguides. Feduccia agrees with Charles Sibley and Jon Alquist, a pair known for their molecular research on the phylogenetic relationships among birds, who conclude that zygodactyly arose independently several times in birds with similar habits. That the same unusual arrangement of the toes turns up in different and only distantly related groups of birds implies that zygodactyly is an invaluable and perhaps essential aspect of the tree-climbing habit.

Does *Archaeopteryx* show any of these additional foot adaptations? *Archaeopteryx* does have a reflexed hallux, an important component of any perching adaptation. However, *Archaeopteryx*'s foot anatomy certainly does not match that of a tree-climber, for *Archaeopteryx* is clearly not zygodactylous and has only a weak flexor tubercle on each of its foot claws. In fact, this tubercle is weaker in the pedal claws than in the manual claws. Therefore, the muscles in the foot simply cannot have been used regularly and powerfully for gripping. There is no other point to which the flexor muscles could attach and still function, nor is there any way these muscles could have been regularly and strongly used without making the flexor tubercle large. The weakness of the flexor tubercle and the lack of zygodactyly makes it difficult to maintain that *Archaeopteryx* was a tree-climbing bird sim-

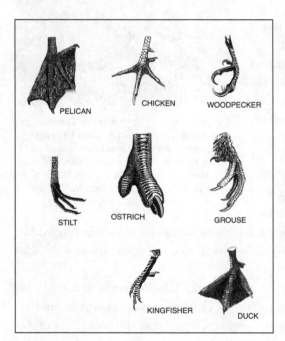

Figure 52. Bird feet are diverse in shape, reflecting the many different habits of living birds. Woodpeckers (upper right) have zygodactylous feet, with two toes pointing forward and two backward. Zygodactyly is typical of many tree-climbing birds, whereas an arrangement with three toes pointing forward and a reflexed hallux pointing backward is more typical of perching birds like the kingfisher (bottom, center) or ground-dwelling birds like the grouse (second row, right).

ilar to a woodpecker or nuthatch. Thus the adaptations in the feet of *Archaeopteryx* give contradictory indications of its behavior. Its foot claw curvature and reflexed hallux are appropriate for perching or climbing, but it certainly lacks the muscle strength and the arrangement of claws necessary for tree-climbing. Reconciling these different indications is difficult. Perhaps *Archaeopteryx* was an occasional but weak and inept climber or percher—a representative of an early and ineffectual stage of climbing and grasping. Most likely of all, perhaps these claws were used in a way not revealed by the current analyses of their shape and function.

Another complicating factor in deducing claw use in *Archaeopteryx* is that it has manual or hand claws as well. No known living bird has manual claws as an adult, which means we know little about the functional interaction of hand and foot claws. In his study, Feduccia also measured the hand claws of *Archaeopteryx* and compared them to the foot claws of living birds of known habits. Even though hands are clearly not feet, this is a valid approach since he was trying to deduce the function of manual claws and not their evolutionary development. Surprisingly, the hand claws of the middle digit of the Teyler, Berlin, and Solnhofen specimens showed a stronger curvature than the foot claws did. These specimens had claw arcs of 155 degrees, 142 degrees, and 145 degrees, respectively, for a mean value of 147.3 degrees, which is close to the mean value for the pedal claws of trunk-climbing birds. A few species of perching birds also had claw arcs in this range. Feduccia explains:

> Most likely, *Archaeopteryx* used the claws of the manus [hand] for clinging to branches because it had not yet achieved the balance that is characteristic of modern birds. It was capable of trunk-climbing but may have done so only occasionally, after flying to the ground. *Archaeopteryx* was probably incapable of taking off from the ground; climbing would have allowed the bird to reach a suitable place from which to launch, where it could take advantage of the cheap energy provided by gravity.

Is this scenario plausible in terms of anything we know about modern birds? The only clue that can be gleaned from living species involves the hoatzin, a primitive South American species that lives in the Amazonian jungles. The hoatzin is a lovely bird, with a bright blue patch encircling each of its large red eyes; a fabulous spiked crest of yellow and blackish feathers adorns its head. Hoatzins are large birds—at about twenty-six inches from head to tail, they are much larger than *Archaeopteryx*—with a yellowish chest, brown striped back, and a long narrow tail (Figure 53). Hoatzins are unusual in several ways, one of which is that hatchling hoatzins have claws on their wings. Although some other birds have vestigial wing claws, the hoatzin is the only species known to have functional wing or hand claws. This feature has made hoatzins difficult to classify. Traditionally, they have been grouped in the order Galliformes, but hoatzins are so aberrant that their affiliations have been debated for decades. Galliforms include all the fowl-like or chicken-like species such as turkeys, grouse, partridges, and pheasants. Most galliforms share a body plan that is generally adapted for terrestrial life, with large feet and claws; their short rounded wings and powerful breast muscles enable them to take off from the ground with a rapid burst. Not so the hoatzin, which is a poor flier and has plumage and markings that re-

Figure 53. The hoatzin is an unusual bird of the South American rain forest. As a fledgling, the hoatzin retains a claw on each of its wings, somewhat like the claws on *Archaeopteryx*. When predators threaten, the young hoatzin uses this claw to clamber out of its nest and drop into the river below. The claw is lost when the primary feathers appear.

semble those of cuckoos. Hoatzins also have a unique digestive system that is different from that of any other known bird. Recent research shows that the hoatzin may not be a galliform after all. Comparisons of the hoatzin's DNA sequence with those of other birds suggests it is much closer to the Cuculidae, the family of cuckoos, than to the galliforms.

In any case, the hoatzin is a largely arboreal, not terrestrial, species that flies clumsily for only short distances. The most astounding thing about hoatzins—and the one that brings them into discussions about *Archaeopteryx*—has to do with their development and strategies for predator avoidance. A hoatzin nest is usually a crude tangle of twigs, placed on a tree branch over a river. The young hatch from the eggs after twenty-eight days. Whereas galliform and cuculid nestlings are physically precocious, walking and feeding within hours of hatching, hoatzin nestlings are altricial, or slow to develop physical maturity. Hatchling hoatzins are nearly naked and helpless; they require intensive parental care in the nest for an unusually long fledgling period. This prolonged period of dependency makes nestling hoatzins especially vulnerable to predator attack. Mostly they huddle quietly in the nest, hoping to escape detection. Their other defense involves clambering out of the nest into the adjacent branches using their feet, bills, and the two clawed fingers on their barely feathered wings. From this position, a young hoatzin will jump out of the tree into the water below, bobbing up again like a cork. The youngster swims awkwardly to the stream's edge and scrambles out and up the trunk again, using its clawed hands and feet. After the hatchling phase is over, the manual claws disappear and the wings and primary feathers develop normally. Although the hand claws of nestling hoatzins are not an evolutionary retention from some *Archaeopteryx*-like ancestor but an independent development, the comparison is irresistible. The similarity and rarity of clawed wings is so striking that the late Pierce Brodkorb, a well-known ornithologist at the Florida State Museum in Gainesville, proposed that the hoatzins were a model for the function of *Archaeopteryx*'s hand claws.

There is one telling difference between *Archaeopteryx* and hoatzins that seriously weakens the analogy: unlike hoatzins, *Archaeopteryx* cannot flex its wing and bend its wrist laterally to tuck the feathers out of the way, exposing the claws for use. When its wings are flexed, *Archaeopteryx*'s free fingers point backward and their claws curve toward the inner surface of its forearm. When its wings are outspread, *Archaeopteryx* can flex its wrist to move the feathers downward and inward and its claws also curve downward and inward, but not sideways. No matter how the wings are positioned and no matter how fully they were flexed or spread, the claws of *Archaeopteryx* cannot be exposed as the most forwardly placed part of the wing. Even with wings folded as much as is possible, *Archaeopteryx* would have had great difficulty using its hand claws for climbing or clambering without damaging the feathers. The same may be true of hoatzins, even though they have the advantage of flexing their wrists laterally. The risk of damaging the primary feathers may explain why hoatzins retard the development of their primary feathers, apparently to prolong the usefulness of their unique strategy for avoiding predators.

There may have been one position in which *Archaeopteryx* could have used its claws for grasping without endangering its feathers. At the bottom of the flight stroke, when the wings are spread and positioned slightly forward and beneath the body, the claw tips would be facing inward and slightly toward the belly. They might then be used to partially encircle a trunk that lay between the wings, aiding *Archaeopteryx* in climbing upward along that trunk, bracing itself with its claws and forelimbs while the hindlimbs walked up the tree like a lineman climbing a telephone pole or a Polynesian islander ascending a coconut tree. To fit between the extended wings, the trunk would have had to be quite slender; the distance between the right and left shoulder joint of *Archaeopteryx* is only 27.6 millimeters (little more than an inch), and the claw-to-claw gap could not have been much wider if *Archaeopteryx* had to exert muscular power for grasping. Happily, the plant fossils at Solnhofen indicate the presence of many shrubs and bushes with trunk diameters no greater than about 1.5 to 2 inches (3.8 to 5 centimeters). This method of climbing works best on an unbranching tree, like a palm or a coconut, rather than on a shrub or bush, where the protrusions, branches, and twigs make climbing awkward and inefficient. Still, such a technique may have given *Archaeopteryx* a way to regain sufficient height if, as Feduccia believes, it was incapable of taking off from the ground.

Feduccia's work is thought-provoking but does not ultimately resolve the function of the wing claws, especially since they do not show the wear and blunting that would be expected if they were regularly used for climbing. The most ingenious proposal about their function comes from Siegfried Rietschel of the Staatliches Museum für Naturkunde in Karlsruhe, Germany. He suggests that the manual claws of *Archaeopteryx* may have been used to groom the feathers. The stunning fact is that grooming is an almost entirely overlooked issue in the evolution of birds, yet it is vital to the success of birds as feathered animals. Preening or grooming keeps the barbs and barbules of the feathers hooked together, which stiffens the flight feathers and enables them to act as airfoils. Air passes through an "unhooked" feather, rather than being diverted around it. Grooming also spreads oil from the preening gland on the tail to the feathers, which makes them water repellent and keeps them from disintegrating into a limp and soggy mass. Finally, grooming also keeps the plumage clean and fluffed up, so that it traps air and insulates the bird from excessive cold or heat. Preening is of tremendous importance to maintaining the health and flying function of birds and is hence one of their most time-consuming behaviors. Any creature with anatomically modern feathers, like *Archaeopteryx,* must have preened extensively.

But how did *Archaeopteryx* groom its feathers? Today, birds preen primarily with their beaks and to a lesser extent with the claws on their feet. *Archaeopteryx* has two toothy jaws that preserve no anatomical indication of an external beak as well. Its mouth was probably lizard-like rather than avian in appearance. *Archaeopteryx* may not have been able to groom its feathers effectively with its teeth, which seem more suitable for combing through the gaps between the barbs and barbules to remove debris and parasites, but it is difficult to see how teeth could hook the barbs and barbules back together again. Rietschel argues that, even if jaw-grooming were feasible, *Archaeopteryx*'s neck was less flexible than that of modern birds, so *Archaeopteryx* may have had trouble simply reaching, picking, fluffing, and rearranging the feathers over most of its body for this reason, too. The long, stiff tail of *Archaeopteryx* probably posed another problem, making the tail feathers much harder to reach than those of modern birds. Rietschel suggests that the long arms of *Archaeopteryx* would have been able to reach many parts of the body, and the sharp-tipped claws may have been used for combing and cleaning the feathers as needed.

A crude analogy can be drawn between the postulated grooming function of *Archaeopteryx*'s claws and the special grooming claws of some mammals. For example, the potto, a nocturnal African prosimian, has a specialized grooming claw on its second toe in contrast to the flat fingernails on all other digits. The claw on the second toe is probably a primitive trait retained from the potto's ancestors because grooming and cleaning the fur is of such enormous importance. Rietschel's hypothesis is entirely consistent with the primitive nature of the wrist in *Archaeopteryx* and the persistence of manual claws. Although there is little direct evidence that *Archaeopteryx* used its manual claws for grooming, it is a suggestion that certainly warrants further investigation. The idea seems plausible and focuses on a major biological need for any feathered species that has been heretofore neglected. And if Rietschel is correct, grooming is an alternative and distinctly nonarboreal explanation for the existence of manual claws in *Archaeopteryx*. The question remains unanswered: how did *Archaeopteryx* rehook the barbs and barbules together after cleaning them? Could it press its lizard-like lips together firmly enough to rehook the barbs and barbules—or did it perhaps use its tongue? We simply do not know.

The heart of the arboreal hypothesis is the proposition that gliding served as an intermediate step between terrestrial locomotion and full flapping flight. From this basic premise are derived a number of important corollaries that are generally incorporated into the theory:

- Early ancestors of birds engaged in a series of locomotor behaviors starting in the trees and involving increasing aerial abilities: simple arboreality, jumping, parachuting, gliding, and eventually flapping.
- The ancestors of birds must have been arboreal because an assist from gravity was needed to make takeoff possible in the early stages of the evolution of flight.
- *Archaeopteryx* is an example of the gliding phase in this evolutionary sequence.
- Since tree-to-tree glides are not always possible, *Archaeopteryx* will show adaptations to climbing that were used to regain height.
- Perching adaptations, used for balancing on branches, were also important in early phases.
- Gliding and perching require that there were trees, shrubs, or other elevated perches available in the habitat of *Archaeopteryx* and other avian ancestors.
- Because the ancestors of birds were arboreal, they are unlikely to have been terrestrial dinosaurs.

Are these aspects of the "trees down" theory supported by the evidence? Yes and no. An evolutionary transition from gliding to flapping flight is demonstrably feasible, having occurred in bats. While feasibility does not prove that this transition also occurred in birds, neither does it refute the hypothesis. Certainly each of the hypothetical stages between terrestriality and full flapping flight occurs among living animals and constitutes a viable adaptation that demonstrably permits takeoff.

Other tests of components of the arboreal hypothesis involving environment and anatomy are more equivocal. Large trees and forests did not grow in the Solnhofen area but smaller, shorter shrubs did and may have sufficed. Anatomically, *Archaeopteryx*'s manual and pedal claws have curvatures suitable for climbing, yet *Archaeopteryx* lacks the other aspects of climbing and perching adaptations, such as the enlarged flexor tubercle on the claw phalanges and zygodactyly. How the manual claws functioned is unclear. Although hoatzins climb or clamber with their wing claws, *Archaeopteryx* lacks the wrist structure that would orient the claws correctly to enable climbing and would protect the feathers from damage while climbing. This wrist structure would seem to be a critical part of the functional complex for climbing with manual claws.

One of the difficulties in evaluating the arboreal hypothesis is that its proponents are not united in their beliefs. The theory itself is less detailed and less coherent than the "ground up" hypothesis, which means the arboreal theory is not as well-developed as an idea but does not mean it is necessarily incorrect. Apart from their adherence to the idea that gliding is the precursor of flight, the arborealists are united primarily by their disbelief in various aspects of the "ground up" theory. They dismiss dinosaurs as avian ancestors yet have no convincing candidate of their own—nor can they agree upon the attributes that this candidate will have when it is found. Increasingly, "trees down" proponents believe that hindlimb bipedalism must have preceded the specialization of the forelimb for flight, but they do not agree when bird ancestors evolved bipedalism, feathers, or thermoregulation. The strongest aspect of the theory—the one least contested and the one that has, in fact, contributed to a growing consensus that *somehow* the avian lineage started in the trees—is that it solves the problem of takeoff.

Does the "ground up" hypothesis fare any better?

Chapter 9.
Dragons Fly

The "ground up" hypothesis was the first to be proposed and has thus been refined and reworked over a longer period of time than the "trees down" hypothesis. The former might equally well be called the "dragons fly" theory, because from the outset, the heart of this theory has been that dinosaurs (whose name means "terrible dragons") are the avian ancestors. Thus the nineteenth-century adherents of Darwinian evolution, like Huxley and Cope, felt *Archaeopteryx* was most likely to have been descended from a bipedal dinosaur, with *Compsognathus* being a favorite model for such an ancestral creature. In 1879, Samuel Williston embellished the idea by proposing that these dinosaur ancestors were cursorial, running animals as well, linking the idea of a dinosaurian ancestry for birds to a hypothesized rapid, bipedal ground locomotion for dinosaurs. Williston's idea was not a well-developed theory so much as a casual remark, and the cursorial hypothesis is most properly attributed to the eccentric Baron Nopsca—whose occupations included both espionage and paleontology—who revived the idea and refined it into a real theory in 1907, and to John Ostrom, who once again revived and revitalized it in the 1970s and later.

Nopsca's great contribution was that he articulated the logic that supported the "ground up" theory by appraising the different types of flying and flying animals. He argued against the widespread assumption that all flying vertebrates (pterosaurs, bats, and birds) had originated and developed flight in a similar manner. In fact, he observed that flying is not a single adaptation; there are many different ways to fly, and their diversity gives hints as to the varied evolutionary development of these adaptations. According to Nopsca, the fundamental difference among different vertebrate fliers is dictated by anatomical equipment. A bird has a feathered arm wing as its primary flying device while a pterosaur or bat has a skin wing that stretches among the fingers and from the forelimb to the hindlimb and, sometimes, tail. Other examples of animals that use patagia are a mixed bag; there are flying squirrels (which do not, of course, truly fly), sugar gliders (a small Australian possum that sucks nectar), and flying lemurs (which are neither flying nor lemurs). Despite their functional equivalence—wings and patagia are clearly analogous—Nopsca asserts in italics that *"from the mechanical standpoint, patagium and feather are two perfectly different organs."* Why? He explains:

> A patagium is a soft flexible membrane and in consequence requires, to be effective, numerous firm radial supports originating from the body that has to be carried, whereas for a series of semirigid but elastic quills one line of attachment is sufficient.
>
> In consequence of this difference, a patagium-flier must always adapt fore and hind limbs and tail to the support of the patagium, whereas in a generalised feathered animal only the feather-supporting element need become affected by violent specialisation.

"Violent" in the last sentence may in fact be a typographical error for "volant," meaning flying: either makes sense, but the implications are slightly different. In any case, Nopsca is correct in his main point. Patagium-bearing species invariably use quadrupedal locomotion on the ground, to go up trees, or to move along horizontal supports like branches. Nopsca maintained that this four-limb adaptation is an obligate consequence of the evolutionary history of such animals, during which both the forelimb and the hindlimb functioned as weight-bearing organs even as the lift-producing, flying mechanism evolved. Both aerial and terrestrial locomotion in such animals requires adept coordination of fore- and hindlimb activities—and sometimes terrestrial locomotion is seriously compromised by the patagium, which can effectively hobble the hindlimbs.

The logic of Nopsca's argument is straightforward. If an ancestral form already possesses the anatomical and neurological mechanisms to coordinate quadrupedal locomotion on the ground, then it is likely to continue a four-limb pattern as it evolves new adaptations to meet the rigorous demands of aerial locomotion. As in the evolutionary pathway followed by bats, walking on four limbs evolves into flying with four limbs. But going from four-limb locomotion to two-limb locomotion, while also changing from a terrestrial to an aerial substrate, may be too complicated to be successful. Is this contention supported by the fossil record? Yes. There are few evolutionary transitions from quadruped to biped, suggesting that this is a difficult evolutionary change to make, and in every bipedal form for which there is a good record, the animal has remained terrestrial as it evolved bipedality. The four-legs-to-two-legs transition has never, as far as we know, involved a dramatic shift in substrate as well. So, Nopsca would argue, any creature that flies with two limbs must be descended from ancestors that had already become bipedal on the ground.

Birds are a prime example. They have evolved a forelimb-only flying mechanism, so at the outset the forelimb must have been no longer essential for weight-bearing on the ground. Consider the formidable problem of takeoff. During takeoff, a bird or birdlike flier uses a bipedal stance, supporting itself on its hindlimbs until sufficient lift is produced by active flapping. Some birds even jump upward into the air, using their hindlimb strength to help lift their body from the ground. Obviously, if the forelimbs are needed to support the body from below, they cannot be simultaneously flapping to support or hang the body in the air from the wings. I suppose an animal could theoretically stand on all fours and leap into the air at takeoff, flapping all four limbs at once, but this seems both ludicrous and unwieldy. In any case, I can find no evidence for an animal that has used a four-legged, hop-and-flap strategy during the earth's history.

Even during takeoff from a raised perch, using the forelimbs for support poses a problem. Birds cannot possibly hold themselves onto the perch (or stabilize themselves on a swaying branch) with both fore- and hindlimbs until the moment of launching without losing control over their initial aerial position and pitch. Nopsca concludes,

> If we . . . now suppose that Birds, before attaining the *Archaeopteryx*-like state, originated from quadrupedal arboreal animals, and only after having learnt to fly became bipedal, it is

difficult to understand why they in general show Dinosaurian affinities, why they did not use both hind and fore limbs to the same extent for flight as they would have done for arboreal loco-motion, why the bones of the pectoral [chest] region and of the wings show more primitive traces than the hind parts of the body, and why they did not develop a patagium. . . .

. . . Birds originated from bipedal Dinosaur-like running forms in which the anterior extremities, on account of flapping movements, gradually turned to wings without thereby affecting terrestrial locomotion. This is also the reason why Birds became dominant over all the rest of their aerial rivals.

Since the time of Nopsca, the fundamental position of adherents of the bipedal-dinosaur-ancestor theory has remained much the same. Bipedalism implies terrestriality, and usually swift terrestriality at that. The real prob-lem with deriving birds from a bipedal, terrestrial dinosaur is the one that initially stymied human fliers: how to achieve takeoff? Lighter-than-air sacs or balloons that solved the problem for early human fliers don't seem feasi-ble among the birds or proto-birds. Another solution, which was being used by the Wright brothers at Kitty Hawk at about the same time that Nopsca was first developing the cursorial theory, is to move along the ground on a mild incline, like a ramp, with wings outspread and at high speeds until suf-ficient lift is generated to leave the ground. For his part, Nopsca envisioned a rapidly running "Proavis" that increased its ground speed by flapping its proto-wings or arms. The reptilian scales on the trailing edge of the proto-wings might, in time, elongate, thus increasing lift, in a spiral of adaptation and advantage that would eventually culminate in the development of feathered wings. Takeoff was simply a matter of running fast enough with outspread arms that had an effective enough airfoil upon them, in Nopsca's view. Certainly feathers and reptilian scales are closely related embryologi-cally, and there is little doubt the former evolved from the latter. The main difficulty is that flapping the proto-wings would create considerable wind resistance, with the result of decreasing speed and inhibiting takeoff, rather than increasing speed.

The Wright brothers devised an alternative and ingenious system that in-creased ground speed effectively. Their device included a large weight, which was dropped from a height and thus provided the power to propel their glider forward, like a rock out of a catapult or slingshot. Since an avian ancestor could

not have made use of a slingshot option to increase ground
speed, then "Proavis" would
have done well to follow an
evolutionary route followed by
many lineages that specialize
in running, such as cheetahs,
horses, or antelopes. Cursorial species like these are
marked by a consistent suite of
adaptations in the limbs, many
of which are found in both
theropod dinosaurs and birds
(Figure 54). Fast-running
species are almost invariably
quadrupedal, for bipedality
generally decreases speed
rather than increasing it.
Compared to their noncursor-
ial relatives, running species
have elongated the segments
of the forelimb below the elbow and the segments of the
hindlimb that lie below the
knee, to increase speed. Most

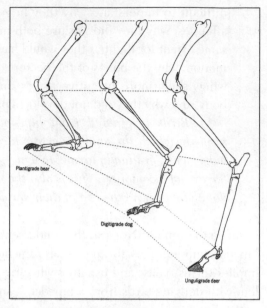

Figure 54. The structure of an animal's hindlimb reflects
its habitual ways of moving. In this diagram, a bear, a
dog, and a deer show the evolutionary changes that
occur with progressively greater adaptations to
cursoriality. The portion of the foot in contact with the
ground is reduced from the entire foot to the last segment of the toes; the number of toes is also reduced.
The segments of the limb below the knee are
elongated, especially the bony heel, which provides
rapid acceleration during pushoff.

of the muscles used for propulsion are moved higher up on the limbs, into the
thigh or upper forelimb, making the lower segments of the limbs little more
than bone covered by tendons and skin. Typically, the number of fingers and
toes is reduced in cursorial species, which have no need to grasp the substrate
upon which they move. In the hindlimb, running species show a characteristic
adaptation that involves the bony process of the calcaneum, the ankle bone that
underlies the heel. This process is where the Achilles tendon of the powerful
gastrocnemius muscle inserts. The gastrocnemius muscle is largely responsible
for the pivoting of the foot around the ankle joint, causing pushoff and rapid acceleration from a standstill. In running species, the bony heel is elongated to
enhance the mechanical action of the gastrocnemius muscle and make it more
effective. Finally, in cursorial animals, the key joints in the limbs are reshaped
to restrict the possible planes of movement at the elbow, wrist, ankle, and knee.

As a result, the limbs are guided into a strict fore-and-aft movement with little or no possibility of side-to-side or lateral movement. This restricted joint movement helps translate all of the muscle power into forward motion. While these anatomical adaptations form a coherent package that says "runner" to any functional anatomist, birds deviate from this profile in one important way: they are bipedal and show cursorial adaptations only in the hindlimb. There are some examples of other cursorial bipeds. These animals first evolved bipedality and then evolved mechanisms to enhance speed; no slow-moving quadruped seems to have evolved adaptations for speed and bipedality simultaneously. If the same is true of birds and theropods, this implies that bipedality preceded cursoriality, which in turn preceded flight.

In the light of these facts, Nopsca's proposed role for the proto-wings of "Proavis" seems nonsensical. The hindlimbs of theropods and *Archaeopteryx* certainly show some of the predicted adaptations for cursoriality, but it is cursoriality based on an already bipedal anatomy. If a quadrupedal animal removed the forelimb from its basic role of providing thrust against the ground and used it as a rather inefficient and poorly shaped airfoil, the result would not have been improved speed. Ostrom adds graciously,

> Perhaps Nopsca was misled by the common habit of many ground birds that flap their wings vigorously during the take-off run. A take-off run is necessary in many ground birds to build up the speed required for flight and the flapping merely provides the necessary lift and propulsion as soon as flight velocity has been achieved. It does not contribute to cursorial acceleration.

Ostrom's updated scenario involves an entirely different use of the arms by "Proavis." He believes *Archaeopteryx* is derived from a small, terrestrial, swiftly running theropod. *Compsognathus, Velociraptor,* and *Deinonychus,* all of which have strong, elongated fingers and toes tipped with sharp claws for catching prey, are often mentioned as models. However, the latter two species are much larger in body size than any reasonable *Archaeopteryx* ancestor. As Larry Martin, who is a man of substantial dimensions, remarked in a television interview, "*Deinonychus* is about my size." *Archaeopteryx*, in contrast, was only about the size of a bluejay. *Deinonychus, Velociraptor,* and *Compsognathus* also lived contemporary with or later than *Archaeopteryx* and thus are extremely poor candidates to be the literal ancestor of *Archaeopteryx*. But Ostrom and his followers have never suggested that *Deinonychus* was the literal ancestor of *Archaeopteryx;* they only propose that it, like the much more diminutive *Compsognathus,* provides some insights into what that ancestor must have been like ecologically and anatomically.

What is important, Ostrom maintains, is that *Deinonychus* and other bipedal raptors show the primary adaptation of the theropod hand, which is for seizing and grasping small prey. This in turn indicates that the forelimb no longer played a supportive or propulsive role. Bipedality was *obligate* among the ancestors of birds, Ostrom emphasizes. Thus *Deinonychus* exhibits an anatomical condition that, in an earlier and smaller species, would have been suitable for an ancestor of *Archaeopteryx.* Dinosaurs like velociraptors with these adaptations are generally classified under the name Maniraptor, or hand predators, and Ostrom believes the hand of *Archaeopteryx* shows similar claws and proportions because its hands, too, were adapted for seizing prey. The key issue is that a hand adapted for seizing prey is not being used for terrestrial locomotion and is thus potentially free to evolve into part of a forelimb-based wing.

Thus the "ground up" theory links the habitat in which flight evolved tightly to other matters. Its adherents also defend a terrestrial, cursorial dinosaur as the ancestor for birds. Simply put, according to this theory, dragons fly—and become birds. And one attribute that makes those dragons good candidates for avian ancestry is their bipedalism, for an essential aspect of the "ground up" theory is that bipedalism preceded flight, so that the forelimbs were freed from other duties to evolve into wings. Another key aspect of the "ground up" hypothesis is that feathers also preceded flight, meaning that their initial function was different from what it is now. It has long been argued by Philip Regal and others that feathers, like fur, may have been fundamentally a structure developed for thermoregulation and insulation that was only later coopted for flight. If feathers are thermoregulatory devices, their evolution before flight evolves implies that warm-bloodedness also preceded flight.

In its simplest form, the "ground up" hypothesis pictures a small, feathered, bipedal dinosaur that runs along the ground rapidly, chasing small prey. In Ostrom's early versions of this scenario, the evolving wings functioned as an insect net, but modeling by Caple, Balda, and Willis indicated that this role for forelimbs would be a greater hindrance than a help. Their research suggested that a theropod pouncing and leaping to catch insects would benefit from extending its already feathered arms to improve balance and steering during its leaps. Any increase in the efficiency of the feathered airfoil would improve stability and maneuverability and would thus be selectively advantageous. Eventually, the feathered forearms would begin to create lift, prolonging the leaps into short-duration, low-altitude flights that improve predatory success without running the risk of catastrophic crashes. As flights were prolonged, the need to flap increased, so that the flights could be lengthened further or to cope with updrafts and crosswinds.

By learning to steer and fly at low altitudes with minimal danger, before powered or flapping flight was initiated, these early flying dragons were blazing a pathway to flight that would later be followed, unwittingly, by the Wright brothers. Their success in developing powered flight, where others failed, was due largely to the fact that they first learned to steer and maneuver while gliding on the wind a few inches above the sands at Kitty Hawk. Takeoff was awkward, dependent upon suitable winds and a slingshot-type device, and crashes were frequent. But a crash meant only repairing the apparatus, not burying the pilot as takeoff from cliffs or other high places sometimes did. Only once the brothers had mastered the techniques of maneuvering in the air did they proceed to incorporate power into their flights. Powered flights provided a more graceful and practical means of taking off and extended the height and duration of flights past the maximum that could be achieved through gliding.

Starting from the premise that flight (and birds) began to evolve from a terrestrial form, there are six significant corollaries.

- *Archaeopteryx* is descended from one of the theropod dinosaurs.
- Theropod dinosaurs were already bipedal and terrestrial.
- Bipedalism precedes the evolution of forelimb-powered flight, because quadrupeds evolving flight involve all four limbs in the wing structure.
- Feathers also precede the origin of avian flight because a bipedal terrestrial bird ancestor can only take off if it already has proto-wings capable of generating lift.
- If insulatory feathers precede the origin of avian flight, so too does warm-bloodedness and a high metabolic rate, which are needed to power flight.
- Taking off from the ground is logically the easier evolutionary pathway to flight because it involves less risk of serious injury, thus permitting more time to master the art of aerial maneuvering before full flapping flight is achieved.

What direct evidence supports the tenets of the cursorial theory?

The first issue is the descent of birds from theropod dinosaurs, specifically the maniraptors. This idea is supported by the many detailed anatomical resemblances between theropods and birds, especially if *Archaeopteryx* is summoned to help as a transitional form. Ostrom's original work in this regard, like that of Thomas Huxley and many before him, was insightful and indeed brilliant—but his methods are perhaps a little old-fashioned and nonquantitative. However, Jacques Gauthier of Yale University, sometimes working with colleagues, has filled the methodological gap nicely. Relying on information about seventeen different taxa, Gauthier identified eighty-four specific anatomical features of Saurischians, the lizard-hipped dinosaurs that include the theropods, and subjected them to rigorous cladistic analysis. The results, published in 1986, link birds firmly to the maniraptors and place them especially close to those like *Deinonychus*. Gauthier's technical analyses are careful and well-reasoned, and they have convinced most paleontologists and anatomists of the correctness of his conclusions. The overwhelmingly dominant view today is that some theropod dinosaur was the ancestor of *Archaeopteryx* and all later birds. Gauthier's 1986 work, in particular, has been influential; a review of avian origins, published in 1991, called this paper "the benchmark to which subsequent studies should refer until it is supplanted by a new, even better corroborated hypothesis." What is particularly telling about this evaluation is that the review's author, Larry Witmer of Ohio University, trained with Larry Martin and has had a thorough exposure to Martin's criticisms of Gauthier's and Ostrom's ideas. Yet Witmer still finds Gauthier's work praiseworthy and convincing. Witmer also lays out the task facing the critics, like Martin, who remain skeptical of these conclusions. He writes:

> Those who disagree with Gauthier's conclusions must (1) discredit Gauthier's analysis by demonstrating *numerous* mistakes in character analysis, and (2) propose a similarly explicit phylogenetic hypothesis incorporating all pertinent data and accounting for Gauthier's characters. It is insufficient to dismiss the entire analysis by showing problems in a few characters or by proposing that a single "complex" character indicates different relationships. [italics in original]

Witmer's challenge to provide a more convincing and analytically rigorous alternative has not yet been met. Of course, the ultimate challenge to the defenders of the "ground up" hypothesis—that of identifying the specific ancestor of birds—also remains unmet, for no one is yet willing or able to designate any particular dinosaur species as directly ancestral to birds. However, this is a more minor problem, for paleontologists can rarely find and recognize the specific ancestor of a later form because fossilization itself is such a rare occurrence.

Gauthier's work, added to Ostrom's less quantitative approach, has demonstrated an anatomical similarity between theropods and *Archaeopteryx*. The question might be asked, Just how similar are they? And the answer comes from the extraordinary history of discovery and recognition of *Archaeopteryx:* they are very, very similar indeed. Not one but two of the seven specimens of *Archaeopteryx*—the ones from Eichstätt and Solnhofen—were actually mistaken for small theropod dinosaurs for many years, until the feather impressions were noticed. The emotional impact of such a mistake occurring not once but twice is strong. The only counterevidence to these anatomical resemblances are the legitimate questions that have been raised, but not yet satisfactorily answered, as to whether the similar features are homologous or analogous. The burden of proof is heavy and no one is offering to shoulder it, because (whatever their true relationship) dinosaurs and birds are separated by a broad evolutionary gap into which many innocently contrived theories are likely to fall to their deaths.

The second issue is that the "ground up" hypothesis designates a bipedal dinosaur as the avian ancestor, which implies strongly that this ancestor was terrestrial. This, too, is a highly reasonable assumption. None of the maniraptors have been suggested to be arboreal and, indeed, many bipedal dinosaurs of all sizes and types have left terrestrial footprints. Strict bipedalism, in which the forelimb plays no locomotor role at all, is primarily a terrestrial adaptation that makes getting up a tree or shrub (using only feet and legs) awkward. Nonetheless, there are some tree-climbing birds that can climb up or down trees using only their hindlimbs, such as woodpeckers and nuthatches. These birds need special adaptations for vertical climbing, like zygodactylous feet that anchor the birds firmly to the tree trunk. As a rule, their feet are also unusually strongly built, with stout bones and powerful muscles for gripping with their toes, as might be expected. Woodpeckers also have strong muscles for bracing their tails against the trunks as they ascend, head first. (In contrast, nuthatches habitually climb head first down trunks and do not brace their tails.) Neither *Archaeopteryx* nor any maniraptor shows adaptations of the foot similar to those found in zygodactylous birds.

211

Another appealing point is that, by postulating an already-bipedal ancestor, the cursorial theory suggests an easily imagined and seemingly natural
evolutionary transition. Speaking teleologically, to make a bipedal,
hindlimb-powered, terrestrial species into one that also has forelimb-
powered flight, you need only graft a flight-adapted forelimb onto the creature. The advantage is that the original locomotor system remains intact
and fully functioning while a second system is added. Also, as Nopsca and
many others since his time have pointed out, the fact that quadrupedal
fliers or gliders are invariably descended from quadrupedal walkers certainly implies that bipedal fliers can evolve only from bipedal walkers.

A fourth major point revolves around the evolution and original function
of feathers. The "ground up" theory requires that the avian ancestor already
had feathers before it evolved flight. Since the original role of feathers is controversial, I wondered if there is other evidence to support the preflight appearance of feathers. Just possibly, the anatomy of *Archaeopteryx* itself
constitutes such evidence. Although *Archaeopteryx* is clearly feathered, it is
not clearly capable of flight. Consider the seven key adaptations to avian-
type, flapping flight: (1) flight feathers (to make a wing out of a forelimb); (2)
hollow bones filled with air sacs; (3) a furcula and a keeled sternum to support powerful flight muscles; (4) anatomically specialized bones of the shoulder to permit a wing flip to be performed; (5) laterally bending wrist joints
that fold the feathers up against the body; (6) special features of the forelimb
joints and musculature that extend or fold the wing as a unit with minimal
muscular effort; and (7) fused tail vertebrae forming a pygostyle that controls
the movements of tail feathers. The only one of these adaptations certainly
possessed by *Archaeopteryx* is feathers. While *Archaeopteryx* has hollow
bones, whether these bones were filled with air sacs is less clear. The third
adaptation is arguably present, for *Archaeopteryx* has a furcula, and the specimen designated as *Archaeopteryx bavarica*—and only this specimen—possesses a bony sternum, which is nonetheless unkeeled. As for the other
adaptations, *Archaeopteryx* has a long tail, not a pygostyle; a dinosaurian
shoulder anatomy incapable of producing a wing flip; and a bony wrist that is
not built to bend laterally and tuck the feathers against the body. Even the
automatic linkage that permits extension and flexion of the entire wing is not
fully formed, although it may exist in incipient form. In short, *Archaeopteryx*
lacks most of the adaptations for full, flapping flight, although there is an
abundance of ideas about what *Archaeopteryx could* do.

Archaeopteryx was almost certainly not an adept flier, despite the near-universal practice of calling it a bird. Birds have feathers, and scientists and laypeople alike cannot resist the temptation to treat this statement as an equivalence. If birds = feathers, then feathers = birds, and *Archaeopteryx* is a bird. But, as Jacques Gauthier cautions, there is a danger in accepting such equivalencies—a danger that constitutes a major drawback to the cladistic methodology that he himself has used to such effect. Inadvertently, the technique tacitly implies that anatomical characteristics held in common actually *define* a cladistic group, like a species or genus. In cladograms (Figure 55), the most primitive or generalized forms appear on the left, with the more advanced or specialized ones placed progressively farther and farther to the right of the diagram. The nodes, or branching points, are usually demarcated by sets of one or more new, evolutionary characters that typify all taxa indicated to the right of that node. This format produces a strong temptation to see the characters placed at these nodes as the defining characteristics of the group to the right. Such an assertion is, as Gauthier observes, antievolutionary; if the traits of all modern members of a group *define* that group, then by definition all members of that group have always had those characteristics, and no evolution can have occurred. A bird born without feathers is nonetheless a bird, and a feathered dinosaur—and the first announcement of such a discovery came as this book

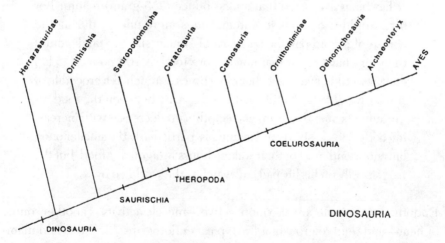

Figure 55. Jacques Gauthier's cladogram groups *Archaeopteryx* and the other birds (Aves) next to the predatory dinosaurs like the velociraptor *Deinonychus*. By convention, the most primitive or generalized forms appear on the left in a cladogram, with the forms becoming progressively more and more specialized on the right-hand side of the diagram. Branching points are defined by the appearance of new, specialized anatomical characters.

(Reprinted with permission from J. Gauthier, "Saurischian monophyly and the origin of birds," in K. Padian, ed., *The Origin of Birds and the Evolution of Flight*, 1986.)

was being written—is not a bird simply because it has feathers. "Characters may aid in recognition of ancestry," says Gauthier, "but they do not define it." In other words, because an animal has feathers, I may recognize it easily as a bird, but that fact does not imply that everything with feathers must be a bird. It is an important point, well worth remembering.

The next point involves the difficult task of deducing facts about the metabolism and physiology of *Archaeopteryx*. Certainly *Archaeopteryx* had feathers; not-so-certainly, *Archaeopteryx* was unable to fly, as it clearly lacked most of the anatomical adaptations linked to flight. The juxtaposition of these two observations implies that feathers therefore evolved to serve a function other than flight, namely thermoregulation. And thermoregulation is wrapped up in a physiological package called endothermy that includes a high metabolic rate, one that fuels high-energy endeavors and an active lifestyle, even in colder conditions. Endothermy, or warm-bloodedness, is usually contrasted with ectothermy, or cold-bloodedness. Chris McGowan, a vertebrate paleontologist at the Royal Ontario Museum in Toronto, has collected some revealing comparisons about food intake among endotherms and ectotherms. He writes:

> It has been estimated that a nine-ounce (255-gram) mammal or bird would use about seventeen times more energy during the course of a mild day than a lizard of similar size . . . [and would] therefore have to eat that much more food. A 100-pound (45 kg) cougar would have to eat about five times as much as a crocodile of similar weight. . . . Notice that the disparity between the food requirements of endotherms and ectotherms decreases with increasing body size. . . . Birds and mammals, particularly the smaller ones, have to spend most of their waking hours foraging for food, but the frog sits idly on his lily pad and watches the world go by.

Endothermic animals are also more active—muscle activity is another source of heat—and tend to grow rapidly. Typical endotherms, like birds and mammals, use a variety of devices—insulation of some sort as well as cooling mechanisms—to keep their body heat approximately stable regardless of the external temperature. Reptiles are often cited as the classic examples of ectothermic animals, at the mercy of the ambient temperature as a main source of heat. They are sluggish when it is cold, can go for long periods without eating, and grow very slowly. The danger in this dichotomy is that it greatly oversimplifies reality. Even cold-blooded creatures regulate their heat closely, by changing their position relative to heat sources and by basking, for example. And some warm-blooded creatures relax their attempts to maintain a constant body temperature; at night, hummingbirds go into a sort of low-temperature torpor, and bears do the same when they hibernate over the winter.

Another issue is overall body size. A large but cold-blooded animal will maintain a fairly constant core body heat despite fluctuations in ambient temperature simply because it takes so long for it to cool down. Galapagos tortoises, which weigh about 450 pounds, lose only about 5°F in body temperature overnight, despite a drop in ambient temperature of 36°F. This strategy has been called "mass homeothermy" or "inertial homeothermy," and may have been a strategy practiced by large dinosaurs.

The question is, What did little dinosaurs do? By "little" dinosaurs, I mean both ones that were small as adults and the young of species that were large as adults. Species like *Compsognathus* were simply too small, even when full-grown, to use mass homeothermy to keep warm. I knew who to ask: Jack Horner, the man who first began finding baby dinosaurs in any quantity. He and his students have been looking for clues to dinosaurian physiology for years, and they have found some interesting evidence. No single piece of evidence constitutes irrefutable proof of dinosaur endothermy, Horner cautions, and valid objections and counterarguments have been raised. Says Horner,

> I certainly do agree with the evidence against endothermy that has been presented to date. Dinosaurs are not warm-blooded just exactly like living mammals and birds. But I do think they are some kind of warm-blooded. I don't think that they received all of their heat from an external source, especially the babies. For the baby dinosaurs to achieve the kind of growth that we see, which appears very similar to modern birds . . . I don't think it can be done without a real elevated metabolism and also some kind of internal energy source. I don't see how it can be done any other way.

And all of the evidence *for* endothermy, taken together, makes a fairly strong case. For example, Horner and his students have studied bone growth in juvenile dinosaurs and have shown that several species of dinosaurs grow very rapidly; hadrosaurs go from small hatchlings to nine-foot-long youngsters in a year's time. No living reptile or cold-blooded species can match that rate of growth, or even come close to it, even under conditions of climate control and force-feeding. Only mammals and birds grow so fast. Then there is evidence of nesting behavior and parental care, mentioned earlier in this book. Horner has shown that hadrosaurs brought food to their babies in the nest, looking after them and protecting them for weeks, maybe months. In addition, there is the direct evidence of an *Oviraptor* fossilized in the action of brooding its eggs—not simply sitting on them but keeping them tucked under its body for warmth and protection. While reptiles lay eggs in crude nests (usually little more than holes in the ground) and a few, such as crocodiles and alligators, may rest their heads on the nest and help open it when the young are hatching, the reptilian commitment to parental care is minimal compared to that among birds and dinosaurs. Parental care does not prove that dinosaurs were endothermic, since rudimentary indications of these behaviors are found among ectothermic crocodiles, for example. But the whole suite of behaviors among dinosaurs—building complex nests, nesting in colonies, hatching dependent babies, and providing food and daily care for those babies over long periods of time—is extremely avian in character. It would be judgmental or anthropomorphic to characterize these behaviors as "advanced," but they certainly show that dinosaurs were much more like birds than they are like living reptiles, and one of the attributes of birds is warm-bloodedness. Finally, there is the testimony of the anatomical adaptations. Many species of predatory dinosaurs are known and their bodies attest to the ability to run fast and capture active prey. Similarly, pterosaurs, which are closely related to dinosaurs, were highly adapted for full flapping flight, a behavior that seems impossible with a reptilian metabolism. (Although reptiles are capable of surprising bursts of anaerobic activity and can actually surpass mammals in terms of the muscle power generated, they cannot maintain a high level of activity for long.) All of the evidence together suggests that some type of endothermy among dinosaurs—and thus in avian ancestors—is very probable.

The final line of evidence that both supports and weakens the "ground up" hypothesis concerns the problem of takeoff from the ground. Logically and practically, taking off from the ground lessens the risk of severe injury to the novice flier, as the Wright brothers found; failures do not involve catastrophic crashes. This is a great strength of the cursorial hypothesis.

However, takeoff from the ground is easy only with the benefit of special equipment. In the case of the Wright brothers, the ultimate solution was a small, lightweight engine; in the case of birds, the "engine" is the bony structure and musculature of the shoulder that make it capable of the wing flip, positioning the wings for the powerful downstroke that produces takeoff. *Ar-chaeopteryx* simply does not have the specialized anatomy of a modern, flighted bird. This fact led Gavin de Beer, in his 1954 monograph on the London specimen, to conclude that *Archaeopteryx* could not fly. He specified three anatomical features of *Archaeopteryx* that led him to this conclusion. First, *Archaeopteryx* lacks a keel, or carina, on its sternum, and this bony ridge is the anchoring point for the pectoral muscles that power the flight stroke. Second, de Beer felt that the short coracoids of *Archaeopteryx*—the coracoid is a part of the bony shoulder girdle—indicated that the pectoral muscle was also short. Finally, de Beer found the bony crest on the humerus, where the pectoralis muscle inserts, to be poorly developed. In sum, these features suggest a small or feebly developed pectoralis muscle that, in combination with the inability to perform a wing flip, would seem to make flapping flight and particularly takeoff from the ground difficult or impossible. De Beer's conclusion that the pectoralis was small and poorly developed has been challenged by later researchers, notably Derik Yalden and Alan Feduccia and Storrs Olson. In particular, Olson and Feduccia argue that the carina is not the primary site of origin of the pectoralis muscle but anchors the supracoracoideus, which performs the wing flip, instead. The pectoralis originated from the furcula and a membrane that extends from the furcula to the coracoid (Figure 56), so the unusually robust furcula of *Archaeopteryx* can be taken as evidence of a well-developed pectoral muscle. Their observations suggest that *Archaeopteryx* could

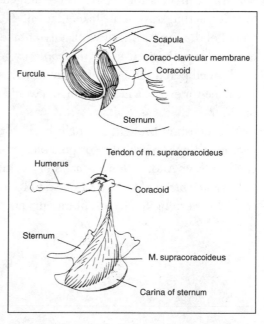

Figure 56. The lack of a keeled sternum in most specimens of *Archaeopteryx* does not mean that this species could not fly, as was once argued. Storrs Olson and Alan Feduccia pointed out that the pectoralis, which powers the downstroke, originates on the furcula and the coraco-clavicular membrane—and the furcula is robust in *Archaeopteryx*. The carina is, instead, the origin of the supracoracoideus, which performs the wing flip for takeoff. Taking off may have been more difficult for *Archaeopteryx* than maintaining horizontal flight.

217

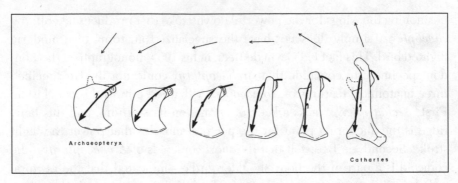

Figure 57. John Ostrom postulated a series of gradual anatomical changes in the avian coracoid and the angle of the tendon of the supracoracoideus muscle that would occur as *Archaeopteryx* evolved to the condition of a modern vulture (*Cathartes*), to make a wing flip possible.

perform a flight stroke, but maintaining horizontal flight once aloft is a less demanding task than takeoff itself. If *Archaeopteryx* is taken as the model of the First Bird, the difficulty of takeoff is the great weakness in the cursorial hypothesis. Could *Archaeopteryx* get off the ground, without using an elevated perch for a gravitational assist? And how did its bipedal, terrestrial ancestor evolve a mechanism that would lift its body off the ground (Figure 57)?

This question brought me right back to the "trees down" hypothesis. Is gliding the only reasonable or possible route to flapping flight, or is there another way? And, if there is another evolutionary route to flapping flight, did *Archaeopteryx* follow it? I decided to look at how flight evolved in the two other vertebrate groups, pterosaurs and bats, for an answer.

218

Chapter 10.

Pathways to the Skies

———

Other than birds, bats and pterosaurs are the only vertebrates that have evolved flight at all. These three groups share some important commonalities that gave them similar starting points in evolving flight. For example, the body plans of their ancestors were probably identical: head, trunk, tail, two forelimbs, two hindlimbs, and five fingers and toes. Vertebrates also share fundamentally similar systems for respiration, circulation, and digestion that must integrate with whatever locomotor system a particular species develops. Too, most bats, birds, and pterosaurs are of the same approximate size range, although some bats and birds are very small indeed and a few pterosaurs were incredibly large. Bats, birds, and pterosaurs thus shared many of the general constraints imposed by body size as they evolved flight. Yet birds, bats, and pterosaurs each found their own pathway to the skies, evolving independently from separate ancestors at separate times. I hoped that by comparing these three flying vertebrate groups and how they evolved flight—insofar as that is known—I would be able to assess the relative likelihood of a terrestrial biped or an arboreal glider as the ancestor of birds.

The exercise of comparing these three groups of flying vertebrates has a dimension that goes beyond anatomy and behavior: time is involved, too. Of the three, pterosaurs evolved first, appearing in the fossil record at about 225 million years and continuing until about 65 million years ago. During their reign as the masters of the skies, pterosaurs speciated and diversified into a wide range of sizes and shapes that reflected differing adaptations. Birds were the next group of fliers to evolve, with *Archaeopteryx* at 150 million years marking the earliest appearance of something that could be called a bird. They coexisted with pterosaurs for many millions of years, but when pterosaurs disappeared at the end of the Cretaceous era and the beginning of the Tertiary—along with many other species, including nonavian dinosaurs—birds persisted. After the disappearance of pterosaurs, birds in their turn began to speciate and diversify in size, shape, and ecological niche. Finally, about 50 million years ago, bats first appeared in the fossil record. Because the earliest known fossil bats are so fully adapted to flight and so similar to modern bats, many paleontologists deduce that the founding members of the bat lineage must have lived

Figure 58. One of the best-preserved specimens of *Pterodactylus kochi* is this exquisite fossil from Solnhofen.

perhaps as much as ten million years earlier still. The sequence of evolutionary explorations of the aerial realm is important. Pterosaurs were the dominant flying animals at the time that "Proavis" and later *Archaeopteryx* made their first tentative forays into aerial locomotion.

What were pterosaurs like? Many pterosaurs have long, narrow, beaky jaws—early ones with teeth and later ones without. Early pterosaurs are informally called the "rhamphorhynchoids," after the genus *Rhamphorhynchus,* and typically have long tails, long fifth toes, and a posture a bit like a dinosaur, with the head protruding in front of the torso (Figure 58). After about 108 million years ago, another type of pterosaur called a pterodactyloid appeared. Pterodactyloids reduced or lost their tails and fifth toes, and some evolved extreme crests that stretch out behind the head in a streamlined fashion. Pterodactyloids typically have a long neck, curved into a birdlike S-shape between the shoulders and the head. By the time birds started to evolve, pterodactyloids were already well-adapted to an aerial niche. As they evolved flight, birds had to compete with pterosaurs for any of the resources that an airborne life might offer—and pterosaurs had already occupied aerial niches for millions of years. What pterosaurs did, how they fed, and how they flew may have limited the options open to birds as they evolved. Similarly, birds formed the ecological context for the evolution of bat flight many millions of years later.

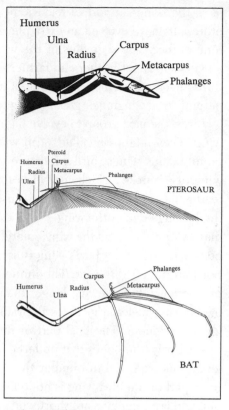

Figure 59. Wings evolved independently in birds (top), pterosaurs (middle), and bats (bottom). Birds have "arm wings" made up primarily of feathers; pterosaurs have "finger wings" in which an elongated fourth finger supports a membranous wing; and bats have "hand wings" in which a leathery membrane stretches between adjacent fingers.

Pterosaurs, birds, and bats did not evolve flight in the same way nor did they share a recent common ancestor. Even a casual observer is struck by the great differences in wing structure among them (Figure 59). The first to evolve, pterosaurs, might be characterized as "finger wings." Their wings are made of a membrane, probably thicker and tougher than a bat's wing, that can be divided anatomically into three parts or segments. The first, shortest part stretches from the torso to the elbow end of the humerus, or upper arm bone. The second, longer part involves the radius and ulna of the forearm and the small bones of the wrist and palm. The third, much longer segment, is the fourth finger, which is greatly elongated and supports the wing membrane to its tip. Not only is each phalanx, the bony support of each joint of the fingers, immensely long but pterosaurs have evolved an extra phalanx than is usual in vertebrates, for a total of four. The segments of the wing are thus progressively longer and longer as they get farther and farther from the trunk. The first, second, and third fingers, though present, are not involved in the membrane but are apparently free clinging or grasping digits; there is no fifth finger.

Birds, the next group to evolve flight, exhibit a different way of making a wing. They might be called "arm wings." Unlike pterosaurs, which have membranous wings, bird wings are composed primarily of feathers; the wing of a plucked chicken reveals few hints of its feathered shape. The feathers are anchored where they insert into the bones of the arm, along the leading edge of the wing; at the trailing edge, the feathers are free and may separate, making the wing emarginated (slotted). As in a pterosaur, the bony skeleton of the bird's wing is arranged in three parts, but the proportions are rather different. The humerus forms the first segment, which is proportionately somewhat longer than in a pterosaur. The middle segment of the wing skeleton includes the radius, ulna, and wrist bones; this is the longest segment. The final part of the wing skeleton comprises the fused metacarpals (the bones that underlie our palm) and fingers so shortened in length and reduced in number that the five-fingered origin of the outermost part of the bird wing is no longer obvious. Thus, in a bird's wing, the first and last segments are short while the middle segment is long.

Bats, the last group to evolve flight, represent yet another evolutionary pathway. Like pterosaurs, bats have a membranous wing. But their membrane is anchored both to the arm bones and to each of the elongated fingers of the entire hand, making bats "hand wings." Thus, while pterosaurs and birds share a similar spar-like arrangement with the arm bones at the leading edge of the wing, bats have a short bony spar terminating in an umbrella-spoke arrangement, with finger bones radiating outward from their small wrist and palm area. Bats also have a three-part wing, but in their case the first two segments are about equal in length, while the outermost segment is the longest. Bats retain all five fingers and use all five to help control the wing membrane. In addition, the trailing edge of the wing membrane is attached to the bat's ankle and often to its tail as well. Movements of the individual fingers, the legs, feet, and tail can all be used to control the tension and camber of the bat's wing. The advantages of this apparently counterbalance the disadvantage of sacrificing the hindlimb as a useful, independent organ for terrestrial locomotion.

A key issue is that flapping flight requires a good deal of energy—so much energy that "sustained flapping flight is clearly beyond the metabolic capacities of modern reptiles" and other ectothermic animals, according to physiologist John Ruben. The two vertebrate fliers that are still extant—birds and bats—are typical endotherms, with high metabolic rates and a steady internal temperature. Were pterosaurs warm-blooded, too? And did they actually flap, or were they gliders? I asked Kevin Padian, an evolutionary biologist at the University of California, Berkeley, who has spent much of his career studying these enigmatic and wonderful creatures. Trained at Yale University as a vertebrate paleontologist, Padian has a keen appreciation of the history of science and has traced the strong influence of old but ill-supported ideas on the popular and scientific interpretation of pterosaurs. His reverence for the history of science has not stopped him from challenging the conventional wisdom about pterosaurs. After more than fifteen years of proselytizing for a new view of pterosaurs, he has not convinced every paleontologist of the accuracy of his ideas but has changed the minds of many.

When Padian began his studies as a graduate student, the view of pterosaurs was detailed and unquestioned. Pterosaurs were obligate soarers: "[T]heir structure is such that it is difficult to understand how they can have had any other means of progression than flying," wrote E. H. Hankin and D. M. S. Watson in 1914. They were envisioned riding the air currents above the waves, fishing with their long beaky jaws; fish remains found in the gut region of some pterosaurs and their overall resemblance in body shape to albatrosses seemed to support this reconstruction. It was even suggested that there was an anatomical locking mechanism for holding the wings outspread and fixed without muscular effort, although closer inspection revealed that this was not so. Pterosaurs were thought to be unable to take off from level ground, as they had none of the shoulder adaptations for performing a wing flip. This tragic inability doomed them to starvation and death if they failed to regain the seaside cliffs from which they had launched themselves. Another aspect of the stereotype was that pterosaurs were almost completely unable to walk on land because, it was thought, the trailing edge of their wing membrane was attached to their legs and tail (if they had a tail, as only some pterosaurs do). The membrane effectively hobbled pterosaurs and prohibited any effective terrestrial locomotion.

Enter Kevin Padian, young and eager to understand pterosaurs. He began his studies by examining the skin impressions of pterosaurs from around the world. The best-preserved specimen is one of *Rhamphorhynchus,* known as the Zittel wing, after K. A. von Zittel, who first described it in 1882 (Figure 60). The specimen shows a clear impression of a narrow wing, somewhat blunted at the tip, and attached to the bony spar of the forelimb along its leading edge.

While the point of attachment of the trailing edge of the wing membrane was obscure in many specimens, what was abundantly clear was that the wing was narrow and streamlined rather than large, broad, and batlike, as was traditionally drawn in pterosaur reconstructions. A particularly fine specimen of *Rhamphorhynchus phyllurus,* eas-

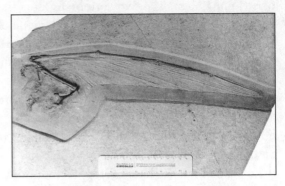

Figure 60. The Zittel wing of the pterosaur *Rhamphorhynchus* preserves the clear impression of a narrow wing membrane.

ily accessible to Padian as a graduate student because it is part of the Yale Peabody Museum collection, showed the wing membrane clearly and invited some calculations. He decided to explore the aerodynamic properties of the traditional and revised reconstructions of pterosaurs, as a test of the feasibility of these opposing views. "All flying animals must obey the laws of aerodynamics," Padian explains. "It seemed to me that the body proportions of pterosaurs could be compared to those of birds and bats, since all three groups must fly under the same aerodynamic principles." Using *Dimorphodon,* a well-known pterosaur from Dorset, England, with a four-foot wingspan, Padian calculated that the traditional reconstruction included a wing area 65 percent larger than would be expected for a flying animal of its estimated body weight. The wingspan was well documented by the fossils and could not be much in error, so the problem had to lie in the breadth given to the wing. To reduce the wing area to an appropriate size without altering the wingspan, however, Padian had to diminish the breadth of the wing until the trailing edge could no longer reach

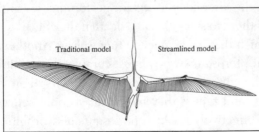

Figure 61. Based on his study of the preserved wing membranes of pterosaurs, Kevin Padian suggested that the wings were streamlined and attached to the torso or knee, not the foot, as older reconstructions had suggested.

224

the hindlimb or ankle. Now the wing membrane stopped at about the level of the pelvis (Figure 61). Puzzled, he returned to the Yale *Rhamphorhynchus* again and looked more carefully; suddenly he saw that the wing membrane passed *under* the hind leg, rather than being attached to it, as he and many others had first thought. More study revealed that neither the Zittel wing nor the many other specimens that have been discovered since 1882 revealed any direct evidence of the wing membrane or patagium attaching to the legs, ankles, or tail at all. After convincing himself of this then-startling conclusion, Padian found that Peter Wellnhofer, an expert on pterosaurs from Munich, had already published a similar, narrow-winged reconstruction based on twenty specimens of German pterosaurs that showed no sign of the wing membrane attaching to the hind leg or ankle. The new look for pterosaurs, independently discovered by Wellnhofer and Padian, involves a sleek-winged creature, shaped roughly like an albatross in large species or a swallow in smaller ones.

Padian was confused by the old stereotype, however. If the bat-like wing on pterosaurs was not supported by the fossil evidence, where had it come from? Looking back through the literature on pterosaurs, Padian found that

Figure 62. From their earliest discovery, pterosaurs were often seen as bat-winged, as in this 1817 drawing by Th. von Soemmerring. Von Soemmerring believed the fossil he had discovered was actually a bat.

pterosaurs had been reconstructed on a bat-like model from some of the earliest accounts. When pterosaur fossils were first discovered in the Solnhofen limestone in 1784, they represented a type of animal that fit into none of the known categories. By virtue of their very existence—not to mention their peculiar anatomy—pterodactyls challenged the existing paradigms of what the world of animals was like. This was a particularly awkward occurrence since evolutionary theory had been neither formulated nor accepted, and the world was regarded as largely static and unchanging. What could these strange creatures represent? The oldest reconstruction in the scientific literature, published in 1817 by the German paleontologist Th. von Soemmerring, is of a juvenile now identified as belonging to the species *Pterodactylus*. Von Soemmerring drew a very bat-like image of the skeleton, with a dotted line representing the wings he correctly deduced were present (Figure 62). Those wings started at the neck, swung outward to the first three fingers and then out to the tip of the elongated fourth finger, then swooped down to attach to the ankles and tail of the animal. There were no traces of the wings actually preserved in this specimen, and the lines were carefully dotted to indicate that they were speculative. However, not only did his drawing look like a bat, von Soemmerring thought the pterosaur *was* a kind of bat. His view was contested by no less a figure than the great French paleontologist and comparative anatomist Baron Georges Cuvier, who asserted that this new creature was a flying reptile as early as 1801. But von Soemmerring's illustration stuck in people's minds and continued to have great influence. In 1836, William Buckland adapted von Soemmerring's drawing for his *Bridgewater Treatises,* which showed pterodactyls gliding, wings outstretched, and clinging to cliff faces with the claws of their first three, short fingers, like upright bats

Figure 63. William Buckland adapted von Soemmerring's drawing of the pterosaur for his *Bridgewater Treatises* in 1836 and perpetuated the bat-winged stereotype.

(Figure 63). The *Bridgewater Treatises* were widely read in England and America and were regarded as the authoritative treatment on extinct species. The image of large-winged, bat-like pterosaurs, with membranes linked to their ankles, was picked up and repeated in many popular nineteenth-century works. Then five-toed "rhamphorhynchoids" were found in England at Lyme Regis. The English comparative anatomist Richard Owen was so convinced that pterosaurs had bat-like wings that he had no trouble seeing the unusual fifth toes "as aids in sustaining the interfemoral or caudofemoral parachute" by which he meant the membranous wing that he believed stretched either from thigh to thigh or between the thighs and the tail. Nearly all of the literature stressed the idea that pterosaurs were bat-like, reptilian creatures. So strong was the impression that pterosaurs were bat-like that direct evidence to the contrary was ignored. In 1882, K. A. von Zittel described some newly discovered pterosaurs with an impression of the wing membrane preserved so clearly that he could draw a pterosaur with narrow wings that shrank to a mere tendon below the waist (Figure 64). The tail was completely free of the membrane. Although this was perhaps the first good evidence of the general wing shape in pterosaurs, the broad-winged, bat-like model persisted, and a uropatagium (membrane between the legs and tail) was drawn and redrawn.

The problem, Padian observes, lies with the observer and not with the evidence. Trying not to imagine what is *not* shown—trying to be truthful, in fact—the artist or scientist is led into the trap of portraying something that is *known* although it is not *shown*. Paraphrasing E. H. Gombrich, who wrote about the fantastical illustrations of exotic animals in the thirteenth to eighteenth centuries, Padian explains the problem beautifully:

> The starting point for the unfamiliar is the familiar . . . ; what an artist believes he sees in an exotic subject merely reflects his own experience. This is why drawings of exotic living animals in those times came out the way they did despite each having been "drawn from life"; why an anatomically incorrect drawing appeared correct to the expectations of a Renaissance artist; and why even the eyes of Dürer's portraits were ophthalmologically inaccurate. . . . If these reconstructions [of exotic animals], which could have been tested against the living originals, prevailed it is no wonder that similar reconstructions of pterosaurs and other extinct animals persisted, despite mounting evidence of their inaccuracies. Theory colors perception, and belief is often stronger than observation.

Added to the bat-like wing morphology was the recognition that pterosaurs were obviously reptilian and not mammalian. The end product was a composite stereotype, a creature bat-like in shape and lizard-like in physiology. Pterosaurs were lazy, cold-blooded creatures with wings that attached to all four limbs. Clearly, such an animal could not have been an active flapper, because flapping flight requires a lot of energy. The simplest way to reconcile the reconstructed wing shape and the presumed physiology was to make pterosaurs pure gliders, creatures incapable of flapping flight. Once Padian saw the overwhelming im-

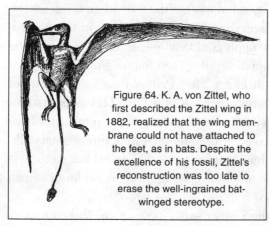

Figure 64. K. A. von Zittel, who first described the Zittel wing in 1882, realized that the wing membrane could not have attached to the feet, as in bats. Despite the excellence of his fossil, Zittel's reconstruction was too late to erase the well-ingrained bat-winged stereotype.

pact of bat-like reconstruction on the scientific interpretation of pterosaurs, he re-examined all aspects of the stereotype. He had already rejected the attachment of the wing to the hindlimb and is still unconvinced by the newly described specimens of *Sordes pilosus,* which are interpreted by some as showing the attachment of the membrane to the ankle.

After the bat-like wing, the next issue was terrestrial locomotion. Bats, with wings that attach to the legs and tail, are hopelessly clumsy on the ground; this is perhaps one reason that they hang suspended upside down when they are not flying. It is not just that the wing membrane makes independent movement of the hindlimbs difficult. Another difficulty in terrestrial locomotion is posed by the orientation of the ball-and-socket joint at the head of the femur or thigh bone. In order for the femur to help control the tension and camber of the wing as it does in bats, it must be splayed out sideways in the same plane as the membrane and the outspread forelimbs. Because this action is so important in bats, their femurs are weak in the weight-bearing position and simply cannot effectively support their body weight when the bat is on all fours. In theory, a bat-winged pterosaur could have made

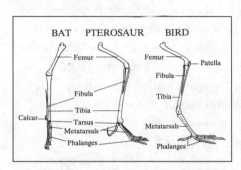

Figure 65. The pterosaur's hindlimb, like that of the bird, is built to support its body in terrestrial bipedality. In contrast, the bat's leg swings out sideways to control the camber of the wing membrane in flight. Because the bat cannot bring its hindlimbs under its body for support, it sprawls awkwardly on the ground.

the same evolutionary compromise, sacrificing terrestrial locomotion for aerial maneuvering and hanging suspended most of the time when it was not flying. However, in reality, pterosaurs are not built like bats in this regard. According to Padian, pterosaurs simply could not hold their femurs out to the side to support their wings without dislocating their femurs from the hip joint. Rather than having femurs built like a bat's, Padian found the femurs of pterosaurs to resemble those of birds instead (Figure 65). This hindlimb structure provides compelling support for his contention that pterosaurs have membrane-free hindlimbs that could be used for terrestrial running and walking. A Padian pterosaur is an animal that has perfected a dual locomotor system—walking bipedally with the hindlimbs, flying with the forelimbs—in an exact parallel to birds (Figure 66).

This brought Padian to the third issue: aerial locomotion. Did pterosaurs glide or could they also fly like birds, with a full flight stroke powered by muscular energy? Padian's careful studies of pterosaur fossils found several anatomical adaptations to full flapping flight. He found that the shoulder joint of a pterosaur is shaped to permit wing movement in several planes. While an up-and-down flapping motion is

Figure 66. In 1983, Kevin Padian reconstructed the pterosaur *Dimorphodon* as a bipedal, running animal—an image in shocking contrast to the traditional bat-winged model.

clearly possible, a twisting movement that brings the wings down and forward—in the type of motion used in slow flying during takeoff and landing—is also possible. The purported locking mechanism in the shoulder proved to be a misinterpretation of a normal bony suture. Still, having the capability for flapping movements at the shoulder is not the same as actually flapping. To find out what pterosaurs *did,* instead of what they seemed to be able to do, Padian examined a more direct indicator of activity: the size of the areas for the attachment of the flight muscles. Big, powerful, frequently used muscles must have large, rugose areas where they attach to the bones that they move; small, weak, and infrequently used muscles have small attachment areas that often leave only barely discernible traces on the bone. Which did pterosaurs have? In a pterosaur, both the sternum and the humerus have substantial areas for the attachment of powerful flight muscles—areas at least as large as those of a comparably sized bat or bird (Figure 67). What's more, a pterosaur has a

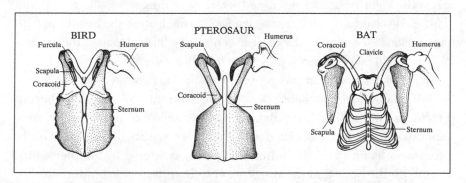

Figure 67. Like a bird, a pterosaur has a large, keeled sternum for the attachment of powerful flight muscles. This similarity of structure suggests that pterosaurs were active flappers, not simply soaring specialists. Pterosaurs may have been warm-blooded animals that dissipated heat via air sacs, like birds, although this conclusion is controversial.

birdlike keeled sternum to provide additional surface area for the attachment of large, powerful flight muscles. Pterosaurs flapped, although the very largest species—such as the monstrous *Quetzalcoatlus,* with a thirty-nine-foot wingspan—must have mostly soared to conserve energy, as large birds do.

Padian found another telling adaptation to flight in the structure of pterosaur bones, one that also challenged the fourth element of the stereo-type: cold-bloodedness. Like birds, pterosaurs have very thin-walled, hollow bones—an obvious adaptation to weight-saving, which is critical to flying animals. In birds, the cavities within the bones are filled with specialized air sacs; these are in turn connected to the respiratory system through tiny holes, or foramina, in the bones. The air sacs are vital to thermoregulation in birds, which face a serious problem of generating too much heat through muscular activity. The air sacs increase the surface area of the respiratory system, which dissipates excess heat efficiently. Their enhanced respiratory system also improves the efficiency with which birds extract oxygen from the air—their rate is almost double that in mammals—and facilitates the exchange of gases, in which carbon dioxide generated by muscular activity is replaced by fresh oxygen from the lungs. Since a good supply of oxygen is essential for aerobic metabolic and energy production, the enhanced respiratory system is a critical element in avian-style endothermy. Pterosaurs not only have hollow bones but also have exactly the same types of foramina. This is strong evidence that they, too, engaged in strenuous flapping activity, consuming high levels of energy and generating excess heat that had to be dissipated through an expanded respiratory system. Perhaps, as in birds, the keeled sternum of pterosaurs also served to keep the intramuscular air sacs from collapsing when the flight muscles were flexed. Pterosaurs thus were able to flap, did flap, and evolved a respiratory mechanism for dissipating the heat from flapping in a manner very similar to that seen among birds. None of these adaptations would be necessary if they were merely passive gliders.

In addition to using an energetically demanding method of locomotion—flapping flight—pterosaurs show other indications of endothermy. Pterosaur bones show the cellular structure typical of animals that grow rapidly, a feature also associated with endothermy. Also, like bats and birds, pterosaurs had a specialized, insulating body covering; in the pterosaurs' case, both the body and the wings seem to have been furred, for hair-like structures are found in specimens with good wing impressions, especially on a species from Kazakhstan known as *Sordes pilosus,* the hairy pterosaur. The possession of true fur or hair is one of the diagnostic characteristics of mammals, and the furry covering on pterosaurs strongly suggests that they, too, were warm-blooded. Not everyone accepts that the structures interpreted as fur on *Sordes pilosus* are fur, but it makes good sense. If pterosaurs had naked skin wings, diurnal flying or gliding over open water would have posed a real risk. I was once told a sad story of a tame fruit bat who was inadvertently left in the sun without shade. Within a few hours, the bat suffered a serious sunburn that caused scarring; this eventually destroyed most of its wing membrane, leaving the bat unable to fly ever afterward. This is a graphic illustration of one of the reasons that many of the animals with skin or membranous wings are nocturnal. Thus, if pterosaurs did not have skin wings and were nonetheless diurnal fliers, they faced a real difficulty. Asked about this problem, Jack Horner quips, "Sunscreen; pterosaurs must have used sunscreen if they didn't have fur." Taken together, the hollow, rapidly growing bones, the air sacs, the fur, and the powerful insertions for flight muscles make a strong case for a warm-blooded, flapping pterosaur. As in birds, many of these adaptations are critical to both flight and thermoregulation.

A spirited defense of the more traditional view was offered by David Unwin of the University of Bristol and Natasha Bakhurina of the Russian Academy of Sciences in 1994. After examining seven specimens of *Sordes pilosus,* they wrote,

> [In these specimens] the main wing membrane . . . is attached to the rear edge of the forelimb, to the body wall from the shoulder to the hip, and to the anterior margin of the hindlimb as far as the ankle. . . . In the absence of evidence to the contrary we conclude that attachment [of the wing membrane] to the hind limb was probably universal in pterosaurs. . . . When grounded, the attachment of patagia to the legs and feet must have severely impeded movement. This, and other evidence for a semi-erect, quadrupedal stance and gait . . . , suggest a poor terrestrial ability and a "gravity-assisted" rather than "ground up" origin of flight for pterosaurs, particularly as the latter requires the ability to run at high speeds.

For reviving this old-fashioned view, Unwin remarked to a reporter, "I find myself with the fuddy-duddies dotted around the world." Nonetheless, he feels that pterosaur hips are not constructed for bipedal walking.

Another objection often raised to Padian's reconstruction is that having a membranous wing that attached either to the trunk or to the upper leg poses a serious aerodynamic problem. How could tension, pitch, and camber be controlled in a membranous wing, supported only on its leading edge by a bony spar? Bats are extremely skilled fliers and manipulate their skin wings elegantly, by moving their bony fingers and legs, for their membrane is stretched between many pairs of bony supports. But Padian-type pterosaurs do not have a series of bony struts (fingers, legs, and toes) with which to control the membrane's shape and tension. Once again, close study of wing impressions—and a careful reading of some of the earliest descriptions of fossils with wing impressions—provided the answer.

What Padian proposes—and he is joined in this by Rayner and Wellnhofer—is that pterosaurs had structural fibers (possibly keratinous) embedded within their wing membranes that controlled their position and tension. These fibers are readily visible in specimens with good wing impressions and were noticed as early as 1882, by von Zittel (Figure 68). Unfortunately, many later workers dismissed them as folds or wrinkles in the membrane, without considering the features of these structures carefully.

First, whatever they are, these elongated structures lie on the ventral or underside of the wing and never cross each other or show abrupt bends, as would wrinkles. Second, these structures are very consistent in width (about 0.05 mm), which suggests a structural origin and not an accidental one. Finally, they are arranged within the wing in a pattern that parallels the long axis of the feather shafts in bird wings, which seems highly unlikely if they represent artifacts of the death postures of pterosaurs. However, if these structural fibers were stiff and only somewhat elastic, they are positioned beautifully to reinforce the supple wing membrane and keep it from billowing and deforming dur-

Figure 68. Striations or fibers in the wing membranes of pterosaurs were first noticed on the Zittel wing, as shown in this close-up. Kevin Padian, Jeremy Rayner, and Peter Wellnhofer propose that these are structural fibers that stiffen and reinforce the free edge of the wing membrane during flight.

ing flight. (Obviously pterosaurs couldn't fly effectively if their wing membranes flapped, billowed, and generally behaved like a flag in the wind.) The existence of these fibers has been widely accepted by other paleontologists after reviewing the evidence. The fibers make the wing a composite material, combining flexibility and stiffness, in such a way as to resolve the main aerodynamic problem faced by a flapping pterosaur. On the whole, the evidence seems strong that at least some pterosaurs fit Padian's model, having narrow wings, flapping flight, and a warm-blooded physiology.

Another challenge to the warm-blooded pterosaur—and dinosaur—comes from physiologist John Ruben of Oregon State University. He is no expert on dinosaurs, as he freely admits, but he is confident in his knowledge of reptilian and avian physiology—and he is

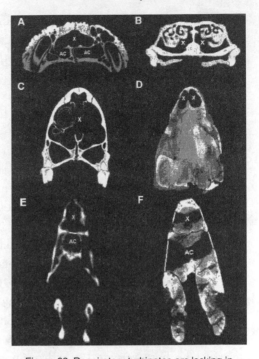

Figure 69. Respiratory turbinates are lacking in cold-blooded animals, like the crocodile (A), but are found in most warm-blooded birds and mammals, like the ostrich (B) and the bighorn sheep (C). The nasal passages of at least some dinosaurs—such as the duckbilled dinosaur, *Hypacrosaurus* (D); the ostrich-like theropod dinosaur, *Ornithomimus* (E); and the tyrannosaurid theropod dinosaur, *Nanotyrannus* (F)—look too small to house turbinates. This observation led John Ruben and colleagues to challenge the idea of dinosaur homeothermy. Each image is a computed-tomography cross-section through the snout or muzzle. X indicates the nasal passage itself.

striving to bring good physiology into paleontological debates. In a series of recent papers, Ruben and others have argued that endothermy in dinosaurs remains unproven and that the most compelling evidence of endothermy will come from the evidence of respiratory turbinates in the skulls of extinct forms. Respiration, Ruben points out, is the fuel line of metabolism; without sufficient oxygen, the metabolism can produce no energy and the animal cannot move, breathe, eat, or reproduce. Respiratory turbinates are a set of fine, elaborately curled bones or cartilages that lie within the nasal cavity of most mammals and birds (Figure 69). They are lined with epithelial cells and function during inhalation and exhalation as heat exchangers, warming cold incoming air or recapturing heat and moisture as animals exhale. Ruben and his colleagues argue that the presence of respiratory turbinates indicates high rates of ventilation and thus high metabolic rates (endothermy) and their absence implies ectothermy. Simply put, endothermic animals must have big noses and big lungs. A preliminary study of four individual dinosaur skulls revealed narrow nasal passages. Similarly, Ruben and Willem Hillenius of the University of California at Los Angeles argue that pterosaurs have chest skeletons too small to support large lungs with avian-style abdominal air sacs and high ventilation rates, a claim that excited strong responses when their data were presented at the 1996 meeting of the Society of Vertebrate Paleontology. In the case of both dinosaurs and pterosaurs, Ruben and his colleagues maintain that respiration rates are more directly and irrevocably tied to metabolic rates than are other, more often cited evidence such as growth rates, locomotor type, or bone growth.

As with any new idea in this field, things are not as simple as they might at first seem. Jack Horner, one of Ruben's coauthors on the study of dinosaur skulls, agrees that the preliminary data are presented correctly but says that he has examined a hadrosaur skull that does have respiratory turbinates—indicating that not all dinosaurs are the same. Horner has thought long and hard about the different types of data about dinosaur metabolism. The issue isn't settled yet. He says,

> What you have to have is not turbinates, but big noses—big chambers to get the air in. We need to ct-scan a whole bunch of dinosaurs to find out what they're like. The data Ruben has presented are right for now, but this is the very beginning of a long, enduring study. It is an interesting avenue to pursue, not an answer.

Gregory Paul, a dinosaur specialist based in Baltimore, also contests Ruben's claim that respiratory turbinates are a "Rosetta Stone" for deducing ancient physiology. Paul observes that respiratory turbinates "can be reliable indicators of respiratory capacity in extinct tetrapods only if their absence or presence is detectable." The difficulty in detecting the presence or absence of turbinates occurs because these structures are, in birds, "extremely variable in placement and form, and are usually cartilaginous," meaning that they leave little or no trace on a fossil. The only viable route, he argues, is to prove that the main nasal passage is inadequate to accommodate functional turbinates—which in turn is problematic, since minimal dimensions for the nasal cavities that still have functional turbinates are not known and probably vary with body size in any case. More argument is likely to be in store as paleontologists and physiologists alike wrestle to resolve how to measure and interpret new information about respiration in pterosaurs, dinosaurs, and *Archaeopteryx*. It is certain, however, that the strong case for pterosaur endothermy made by Padian and others will be reexamined and reevaluated in the years to come.

From what sort of ancestor did pterosaurs evolve? There are no generally accepted fossil candidates for the ancestor of pterosaurs, so the answer depends on the reconstruction of the pterosaur wing and flight apparatus. Those who endorse a bat-winged pterosaur with a wing attached to the leg or ankle generally envision the ancestral form to be a reptilian glider. Pennycuick uses the giant red Asian flying squirrel, *Petaurista petaurista,* to illustrate how he thinks pterosaurs may have evolved. *Petaurista* is one of several gliders that have evolved a cartilaginous spur projecting laterally from the wrist; this spur supports the patagium and allows it to be larger than if it were attached only to the fingers. Although pterosaurs did not (so far as is known) evolve a wrist spur, they did elongate their fourth finger, which would have much the same effect in supporting a larger patagium to enhance gliding. The main requirement for going from enhanced gliding to flapping is simply making the patagium large enough.

A radical variant on this hypothesis of gliding origins is proposed by Dietrich Schaller, who suggests a four-limbed, gliding ancestor for a Padian-type pterosaur. Schaller hypothesizes that pterosaurs arose from a flying-squirrel-type glider, which then evolved into a forelimb-dominant flapping form while still retaining all four limbs in the wing membrane. He cites *Sordes pilosus* as an example of this stage. The next step involved further developing the outermost part of the wing by elongating the fourth finger still further. At this point, the membrane's attachment to the legs disappears as the nature of the wing membrane changes, to become self-cambering, self-supporting, and narrow in outline. At present, no fossil evidence offers direct support for Schaller's idea, and the hypothetical transition from wings involving four legs to wings involving only the forelimbs seems improbable.

235

Proponents of a narrow-winged pterosaur are often the same individuals who support a terrestrial, theropod ancestry for birds. They suggest that the ancestor of pterosaurs was a bipedal, terrestrial reptile that developed wings by enlarging the skin connecting its forelimb to its trunk into a wing membrane. They construct an evolutionary transition from ground-dwelling biped to flapping flier that parallels the one proposed for birds. Indeed, this is the only plausible type of scenario that can be imagined for a Padian-type pterosaur. Thus the debate over the evolution of flight in pterosaurs is structured as a dichotomy—almost the very same dichotomy, in fact, as over the origin of birds.

Does the evolutionary history of bats offer any insights that might be useful in resolving these debates? One advantage is that the adaptations, behaviors, and ancestry of bats are easier to analyze than those of pterosaurs, because there are so many living examples. All living bats are classified in the order Chiroptera, which is subdivided into two main suborders among living bats: the Microchiroptera, which are smaller-bodied, usually insectivorous, and use echolocation to find their prey; and the Megachiroptera, the larger-bodied fruit bats that rely more heavily on eyesight. Both types of bats have wings comprising a nearly hairless membrane that stretches from the hindlimb and body to the immensely elongated hand and between the adjacent pairs of fingers on that hand. The first finger (digit I) or thumb is freed from the membrane and is used for crawling or hanging; digits II through V serve as struts to strengthen and control the membrane, somewhat like the spokes on an umbrella. The hindlegs are also embedded in the membrane and are splayed out sideways to help control the membrane during flight. Unfortunately, the combination of membrane and splayed hindlegs makes bats fairly helpless on the ground. The bat's tail is also embedded in a membrane, known as the uropatagium, which varies in extent from species to species but contributes to the total surface area of the airfoil when it is present. Unfortunately, early bats have left an even poorer fossil record than early birds. The oldest known bat fossil is a complete skeleton of a bat called *Icaronycteris* from the Green River Formation of Wyoming, which is roughly 50 million years old. *Icaronycteris* is generally identified as a microchiropteran; it is clearly capable of full, flapping flight and it already shows skull features linked to echolocation. Thus the evolutionary transition from flightlessness to flight in bats is completely undocumented in the fossil record; there are no clues to guide researchers other than the characteristics of living forms and the constraints of aerodynamics and ecology.

Nonetheless, there is "overwhelming" evidence that bats evolved from an arboreal, mammalian glider. In bats, both fore- and hindlimbs are involved in the patagium, as in gliding mammals. Like gliding mammals, bats have elongated all four limbs to support their wings, although bats have carried this process farther than gliders. Furthermore, gliding mammals are often small and nocturnal, as are bats. Probably these ecological attributes were also part of the ancestral adaptation. Finally, the closest living group to bats is dermopterans, the group represented by the colugo or erroneously named flying lemur; dermopterans are gliders and also hang upside down when not flying. Hanging as a means of resting or perching may also be an adaptation derived from the common ancestor of bats and other gliders. Hanging is so important that many species have evolved a passive anatomical mechanism that locks their feet onto their roost. This mechanism allows them to conserve energy while hanging, which seems to be especially important in hibernating species that are on a tight energy budget.

An important difference between bats, on the one hand, and birds and pterosaurs, on the other, is that bats have not developed an air-sacs-and-hollow-bones cooling system. The reason is that bats do not need an enhanced respiratory system for cooling, despite being flapping fliers. For the same reason, bats also lack a keeled sternum, even though their flight muscles are large. Because the ancestor of bats was a mammalian glider, then it was already endothermic and fully capable of fueling high-energy activities. Evolving flight did not necessitate a wholesale revision of the respiratory and physiological systems in addition to the locomotor one. Thermoregulation in bats is easier because they fly almost exclusively at night, when ambient temperatures are generally lower than body temperature. Convection working on the large surface area of their skin wings (which have little fur) is sufficient to dissipate the muscular heat of flapping.

Bats today are specialized into a variety of ecological niches, which have distinctive consequences for their morphology and lifestyle. Still, bats are more restricted ecologically than birds. In the first place, virtually all bats are nocturnal and the most primitive ones seem to be insectivorous. It seems likely that bats specialized in nocturnal flying and insect catching to invade an unexploited ecological niche, one in which they would experience little direct competition from birds. When bats began to evolve, birds were already adept daytime fliers specialized to fill many different aerial niches. At that time, birds may have fed in the air (a relatively rare adaptation today) or perched on trees and shrubs, taking a variety of food types such as insects and/or fruit. Other contemporary mammals, like true and flying squirrels, already occupied the trees, jumping or gliding from tree to tree, and concentrated on plant foods. This left night flying and insect eating, aided by acoustic not visual clues, as an evolutionary opening for bats. From insect eating, bats may have evolved into nectar lapping and then to frugivory, with some specialized lineages evolving further to take up fishing, carnivory, or blood-sucking lifestyles.

There are several predictable consequences of the original chiropteran adaptations that are still evident among bats today, such as their strongly restricted size range. Among living species, bats range from about 1.9 grams to 1.5 kilograms (0.07 ounce to 3.3 pounds) in weight, the largest being the Old World fruit bats known as flying foxes. In contrast, living, flighted birds vary from a 1.5-gram (0.05-ounce) hummingbird to condors and albatrosses that weigh 12–15 kilograms (26–33 pounds). Although it is sometimes said that pelicans, condors, and albatrosses must be near the upper limit for horizontal flight, larger flying birds once existed. The extinct New World teratorns included a species called *Argentavis magnificens* that had a 6-meter (20-foot) wingspan and a body weight of 80 kilograms (176 pounds). Only after pterosaurs had become extinct

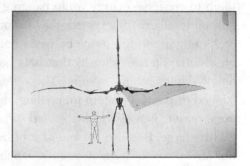

Figure 70. *Quetzalcoatlus northropi* is the largest pterosaur yet found and may have been the largest flying creature that ever lived. Its discoverer, Wann Langston, Jr., reconstructs its wingspan as about 39 feet and its body weight as about 280 pounds. A human figure is provided for scale.

did such avian monsters evolve, however. For their part, pterosaurs had the widest range of body sizes, from small species about the size of a sparrow to the largest, *Quetzalcoatlus northropi* from west Texas, which had an estimated wingspan of about 12 meters (39 feet) and a weight of roughly 127 kilograms (280 pounds). *Quetzalcoatlus* was the largest flying animal that ever lived and must have spent a great deal of its time soaring on thermals or air currents above the waves, rather than flapping (Figure 70). Since the mechanics of flight obviously do not prohibit the evolution of large-bodied species, why are bats so small relative to birds and pterosaurs? Part of the answer is clearly ecological, for bats fly only at night. There are two main reasons that bats are nocturnal. First, they faced competition from diurnal birds, who were already a well-established group when bats were first evolving. Second, bats evolved from a gliding ancestor who was probably nocturnal, as many gliders are. However, flying at night eliminates the possibility of soaring or intermittent flight strategies to save energy because there are no thermals at night. Nocturnality alone thus helps to restrict bats from evolving a larger size.

Bats also lack another adaptation that helps large birds fly: slotted wings. Probably bats haven't evolved slotted wings because their wings are membranous and an unsupported strip of membrane would be an ineffective airfoil (although ways around the problem can be imagined). Among very large birds, the tips of the primary feathers separate, producing slots; this is a typical and extremely useful adaptation. Although the aerodynamic consequence of adding slots to a wing is complex, the net result is obvious: it decreases the energetic cost of flying by increasing lift, reducing drag, and increasing longitudinal stability. Slotted wings can be regarded as a special adaptation of birds, one that comes along with evolving feathers. However, slotted wings cannot be essential to flapping flight among large-bodied species because large pterosaurs lack slots. Perhaps the rule for evolving large body size among fliers is that either soaring or slots—or, best of all, both—are essential.

Bats are restricted by their ecology not only in body size but in wing shape, too. Most bats have aspect ratios between 5 and 6, with the highest values (between 8 and 9) found in molossids; the range in birds is from about 4.4 to 20. Bats generally have larger wings (i.e., wings with a larger surface area) than comparably sized birds or pterosaurs. As a result, bats have consistently lower wing loadings than the other fliers and do not achieve such extreme aspect ratios as do some birds and many pterosaurs. The long, narrow wings and high aspect ratios characterized as adaptations for soaring are simply not found among bats, since this type of flying is not an option for them. Birds that fill a bat-type niche, catching insects on the wing, have a typically bat-like wing loading and aspect ratio. The difference is that such birds have low values of wing loading and average values of aspect ratio by bird standards, which makes such birds relatively slow fliers of high agility, whereas bats with the equivalent build are some of the fastest and least agile fliers among bats, sporting high wing loadings and aspect ratios.

Where do pterosaurs fall in terms of aerial performance? Few pterosaurs are sufficiently well known to be compared to birds and bats in terms of wing shape, but they seem to have been reasonably good fliers. *Pteranodon* was the subject of a research project carried out by a paleontologist, Cherrie Bramwell, and an aeronautical engineer, G. R. Whitfield. Writing in 1974, before Wellnhofer and Padian proposed the streamlined reconstruction of pterosaurs, Bramwell and Whitfield used a traditional, bat-winged reconstruction of *Pteranodon* in their calculations. With a five-foot wingspan, they estimated that *Pteranodon* had a wing loading of 0.36 g/cm^2 and an aspect ratio of 10.5, values similar to those for a frigate bird. In 1983, James Brower repeated their calculations with a Padian-type *Pteranodon* and obtained a higher wing loading of 0.59 g/cm^2, about the same as in a pigeon, and a very high aspect ratio of 19.0, comparable to an albatross's aspect ratio. Both studies concluded that pterosaurs were specialized for low-speed flying combined with great maneuverability. While *Pteranodon* probably could not have flapped powerfully enough or long enough to make progress against a strong headwind, an animal built like *Pteranodon* could hardly help taking off in even a light breeze. With its low stalling speed, high aspect ratio, and moderate wing loading, *Pteranodon* would have only had to face into the wind with extended wings to become airborne, according to Chris McGowan. Furthermore, the stalling speed of *Pteranodon* was so low that the danger of landing awkwardly was greatly lessened. Similarly, when Marden reviewed the demands of takeoff and sustained horizontal flight, he found at least one pterosaur to be an able flier. Using the highest estimate of body weight (550 pounds), Marden calculated that a *Quetzalcoatlus* would need to generate only as much power as a 2.2-pound (1-kilogram) vulture or a 22-pound (10-kilogram) swan to take off. Because takeoff requires more energy than the swan or the vulture can produce aerobically, they rely on short periods of anaerobic metabolism to generate a significant burst of energy, but even modern, endothermic birds cannot sustain that rate of energy production for long. To sustain horizontal flight, *Quetzalcoatlus* would have to produce about as much energy as a swan, which during migration can flap for a long period of time.

The apparent restrictions on the size of bats and the shapes of their wings might be taken as evidence that they are poor fliers—a third evolutionary "try" at evolving flight that wasn't very successful. The contrary is more likely to be true. Bats are better, more specialized, and more committed fliers than either birds or pterosaurs. Bats fly continuously under their own power and are consummate flying performers, capable of hovering, diving, twisting, turning, and flying rapidly even in obstructed areas. Those bats with sonar have clearly added a second powerful and complex sensory adaptation to make them even more effective at their sort of flight. Unlike birds, bats generally feed on the wing, the notorious exception being vampire bats that land on their hosts and lap blood; bats do little else except rest, breed, and nurse young while perched, which consists of hanging upside down.

In contrast, birds have retained many abilities and behaviors that do not involve flight. Most birds obtain their food and consume it while walking, standing, or perching, and in addition carry out their reproductive activities on the ground or in a nest. Relatively few kinds of birds—flycatchers, peregrines, fish eagles, hawks—catch live prey on the wing, and even these land to eat. Terrestrial locomotion is important to birds in a way that it is not to bats, and birds generally have long, strong, and mobile legs. Pterosaurs represent an intermediate adaptation. They are neither so deeply committed to flight, anatomically, as bats, nor probably as adept on the ground as most birds. Pterosaurs almost certainly caught their prey on the wing and may well have consumed it there. Their wings, insofar as their shapes can be accurately reconstructed, are streamlined and narrow, suggesting that they flew much more like a gull-winged bird than a bat. Certainly the largest pterosaurs cannot possibly have flapped continuously as bats do and must have relied heavily on soaring to save energy.

From an evolutionary perspective, the comparison of bats, birds, and pterosaurs reveals some fascinating facts. Pterosaurs, as the first to evolve flight, were able to diversify into a wider range of sizes than either bats or birds would achieve, although most of this diversity appeared after birds had evolved. Pterosaurs seemed to specialize in a narrow-winged shape with a high aspect ratio that was suited for flying in uncluttered habitats and open spaces. They were probably warm-blooded and were certainly active, diurnal fliers that may have been largely or exclusively adapted to foraging over water, probably for fish. After 70 million years, the first bird, *Archaeopteryx*, appeared, marking the beginning of the avian lineage's evolution of flight. Whether its ancestry was dinosaurian or crocodilian, *Archaeopteryx* evolved from an already-bipedal ancestor, one who had freed its forelimbs from other uses. Thus one specialty of the avian lineage is the dual locomotion system. Another key adaptation was the appearance of the feather, a unique structure that may well have arisen for thermoregulation but quickly became the keystone of an avian-style mode of aerial locomotion. Finally, like pterosaurs, early birds seem to have been both predatory and diurnal and closely tied by ecology to water. About 65 million years ago, pterosaurs went extinct, one of the last species being the giant of the skies, *Quetzalcoatlus*. But birds persisted, and by 50 million years ago, fully formed bats appear in the fossil record. The earliest bats, or their proto-bat ancestors, may well have coexisted with the last pterosaurs; they were certainly influenced by the presence of birds, who dominated the diurnal aerial niche by that time. Bats undoubtedly evolved from a different kind of ancestor: a furry, nocturnal, warm-blooded mammal that glided from tree to tree. This ancestry directed them along yet another pathway to the skies. Unlike birds or pterosaurs, proto-bats had no need to evolve a new locomotor system involving only their forelimbs, nor did they need special adaptations to keep warm or cool, since their ancestors had already mastered endothermy. Instead, bats were free to specialize in fast, continuous flight, emphasizing great maneuverability to catch insects. Bats were from the outset, and largely remain, enhanced gliding mammals.

Comparing the evolution of flight in pterosaurs and bats reveals the intimate association among thermoregulation, respiration, energy production, and flight. Evolving from cold-blooded ancestors, birds, insects, and pterosaurs faced one set of issues; bats, evolving from a relatively recent mammalian stock that was already endothermic, faced another. To all, being able to keep warm or cool enough is an important advantage—but bat ancestors had already solved the metabolic problem before starting to fly. Birds and pterosaurs had to evolve several different adaptations either to bump up their metabolic rates, making energy more readily available, or to conserve the energy that was generated, by insulating the organism and titrating heat loss or gain closely. The more efficiently an organism can use its energy, the more successful it will be.

Another strong influence on the way each of these groups evolved flight is the habitual locomotion of the ancestral form. Clearly, birds represent the end result of the evolution of flight from a bipedal ancestor, whereas bats represent the end result of the evolution of flight from a quadrupedal, arboreal (and gliding) ancestor. The evolution of flight in pterosaurs remains equivocal. They may have paralleled the avian lineage, as a sort of featherless bird, or they may have been an odd sort of large, daytime bat. The majority view among paleontologists is at present that pterosaurs were narrow-winged forelimb fliers, in which case they prove that a terrestrial biped can evolve flight—and that all vertebrate fliers need not pass through a gliding phase. But scientific facts are established by evidence, not majority votes, and more evidence is sorely needed.

In the last analysis, flapping flight is a high-risk, high-cost adaptation. It offers some very special advantages: new resources, new means of escaping predators, new areas for nesting and raising young. Still, flight is an exacting evolutionary choice. Here invertebrates and vertebrates part company. For an invertebrate, evolving wings does not mean sacrificing a forelimb that is already useful and functional; insects simply add another appendage. But for vertebrates, gaining a wing is losing a forelimb. When the forelimbs become wings, they become so highly specialized that their ability to function in other tasks is compromised. If the hindlimbs are also involved in the wing, they, too, lose their usefulness for other tasks. In addition to the metabolism being geared up to produce more energy, the respiratory system, too, may be forced to evolve in order to accommodate the new and demanding mode of locomotion. In short, many parts and systems of the body may ultimately need evolutionary revision or replacement if flapping flight is to evolve.

Evolving flight seems enormously complicated because so many systems are involved, so many functions are shifted or compromised. On the face of it, flapping flight seems a foolish evolutionary choice, doomed to failure. And yet it occurred, not once but three times in vertebrates, by groups that followed three different pathways to the skies from three different starting points. And the duration and diversity of the bat, pterosaur, and bird lineages is compelling testimony of the great advantage offered by flight.

Because of that advantage, somewhere in the gaps in the fossil record between reptiles and *Archaeopteryx*, some remarkable evolutionary changes occurred. A lineage that on present evidence seems most probably to have been a small, predatory, endothermic, and bipedal dinosaur began to evolve the mechanisms for flight: wings, feathers, an enhanced respiratory system, hollow bones, and, eventually, a remodeled shoulder and arm.

We do not yet know how this happened in any detail, or why. But by taking a good, hard look at *Archaeopteryx*'s anatomy and adaptations, I thought I could discover how far along the evolutionary trajectory to flight *Archaeopteryx* had come.

Chapter 11.

Flying High

The basic question is, Could *Archaeopteryx* fly?

In trying to answer that question, I was in a sense testing the two hypotheses of the origin of avian flight, for each theory makes predictions about the abilities of *Archaeopteryx*. However, the theories are not symmetrical. The logical heart of the "ground up" hypothesis is that the ancestor of birds, and of *Archaeopteryx*, was a theropod dinosaur. All the elements of the theory—bipedality preceding flight, feathers preceding flight, the development of a forelimb-only wing, and takeoff from the ground—hang on the premise about avian ancestry. In contrast, the vital center of the "trees down" hypothesis is that gliding is a halfway step on the transition to flapping flight. Since gliders are arboreal, the ancestor of birds must have been arboreal and feathers must have originated as a means of building a gliding apparatus. What's more, the ancestor of birds cannot have been among the theropod dinosaurs, because there is no obvious way to get a bipedal dinosaur up a tree. Thus, whereas the "ground up" hypothesis is based on a belief about phylogenetic relationships, the "trees down" hypothesis is built upon an idea about locomotor mechanisms.

In its baldest form, the dichotomy between these theories is expressed by the question, Was *Archaeopteryx* an evolved theropod or did it glide? Phrased in this way, the asymmetry of the alternatives sounds like the setup for a joke—"That was no theropod; that was my husband," or some such. But the question is perfectly serious, and appreciating the asymmetry is the key to understanding the intractability of the debates. Disproving one theory does not necessarily prove the other. The proponents of the two theories start from utterly different premises and proceed to speak of subjects that are literally different, like the proverbial apples and oranges. Small wonder they cannot agree.

As always, *Archaeopteryx* is the focus of the debate. If *Archaeopteryx* could glide but not flap, then it was almost certainly arboreal in habit (with the proviso that "arboreal" is taken loosely). The arboreal hypothesis would also be supported if *Archaeopteryx* could flap but not take off from the ground under its own power, implying that takeoff had to be gravity-assisted. On the other hand, if *Archaeopteryx* could demonstrably take off from the ground and flap, then there is no need to postulate a habit of jumping out of trees. This would, by default, support the "ground up" hypothesis.

A key issue in predicting the aerial potential of any species is its body weight, or mass. Quite simply, the fundamental prerequisite of successful takeoff is that sufficient upward force or lift (in a special sense) be produced to move the body weight from its supporting substrate into the air and sustain it there in some reasonably predictable fashion. Colin Pennycuick believes that body mass is a primary constraint on flying. As evidence, he points out the change in the abundance of flying organisms in different weight classes. Among living species, the maximum weight is about 10 kilograms (22 pounds), the size of swans or albatrosses, and the minimum is about 1 gram (0.04 ounce), the size of the smallest hummingbird. But the abundance of flying creatures is not equally distributed across this broad range of sizes. At the very top of the weight range, there are few species of flying organisms even if extinct forms like the teratorns and giant pterosaurs are included; all of these enormous flying animals have extraordinary adaptations to enable them to fly. Flying species are less rare in the range between 10 kilograms and 1 kilogram (22 and 2.2 pounds) but are still usually strongly specialized in terms of wing shape or habits. Flying animals between 1 kilogram and 100 grams (2.2 pounds and about 0.22 pound) are more numerous and apparently do not require special adaptations beyond those directly associated with flapping flight. *Archaeopteryx*, generally thought to be about the size of a bluejay, falls in this range. But the optimal size class is between 100 grams and 10 grams (about 0.22 pound and 0.022 pound), a size class in which flying animals of many different types with many different adaptations are abundant. Below this optimal range, flying species again become uncommon, with a minimum body weight for vertebrate fliers of about 1 gram (0.0022 pound). Such tiny animals cannot contract their muscles fast enough to stay aloft. (Insect physiology is different, permitting faster muscle contractions; hence many flying insects weigh less than 1 gram.)

According to Pennycuick,

> A nonflying animal, destined to evolve the power of flight, will have some structures that are not yet wings, and do not, initially, perform well as such. To have a high chance of developing flight, [this creature] must be in a region of high potential diversity for flying animals. A nonflying animal over 1 kg is very unlikely to evolve flight, because only a limited range of specialized animals can fly successfully at such large sizes. To have the best chance of success, the ancestor should be in the mass range 10–100 g, where oscillating maladapted limbs at inappropriate frequencies can still produce acceptable results.

This is true because only a few substances—like skin, bone, fat, tendon, muscle, and sometimes feathers—are available for making flying organisms, and these substances have fixed properties of density, tensile strength, compressive strength, and so on. The properties of these materials impose real constraints on the design options open to such animals. All organisms must make their wings big enough (and efficient enough in shape) to produce enough lift to counteract the frictional drag that moves them earthward, without increasing their body mass so much that flight is impossible. In the optimal weight range, meeting the demands of sustained flight with the anatomical materials available is apparently not terribly difficult. Unlike aeronautical engineers, however, organisms cannot invent a new, lightweight substance that they can substitute for bone, for example, in order to attain a larger size. This means that a large bipedal dinosaur—or even a person-sized one like *Deinonychus*—is very unlikely to be directly ancestral to any flying form; Pennycuick argues convincingly that it is to smaller forms that we must look to find the true ancestor of birds. Interestingly, although *Archaeopteryx* itself is small and weighed perhaps 250 grams (less than 9 ounces), it is probably larger than the optimal size for evolving flight.

With a good estimate of body mass, only a few additional pieces of information are needed to make a general prediction about the flight ability of

Archaeopteryx or any other organism. One important parameter is the total wing or airfoil area, a simple measure of the size of the two wings. Wing area is not generally measured strictly accurately but is estimated as the product of wing breadth multiplied by wing length. Since the wings are the main mechanism for producing lift, their size is obviously important. A related ratio, wing loading, reflects the mass or weight (in grams) that must be supported by the lift produced by each square centimeter of wing area. Wing loading directly influences the type of aerial locomotion that is possible for an organism. Animals with very low wing loadings can parachute, if their body weight is low (below about 8 grams), and glide readily if they are somewhat larger. Soaring birds, like lammergeiers or battaleur eagles, typically have low wing loadings. So does the rufous-tailed hummingbird—one of the few birds that can hover—which weighs only 6 grams (0.21 ounce) and has a wing-loading ratio of 0.4 g/cm^2. In fact, animals with low wing loadings fly best at low speeds and are more maneuverable than those with high wing loadings. Rayner estimates the minimum wing loading for flapping flight to be about 0.015 g/cm^2, the measured wing loading for the noctule, a relatively large microbat. At the other end of the scale, animals with high wing loadings—such as ducks—seem to be specialized for fast, powerful flight. Logically, if wing loading is too high, an animal cannot become airborne. The highest reported wing loading for a flying animal is 2.45 g/cm^2, and F. A. Humphrey and Brad Livezey have suggested that 2.5 is the threshold wing loading above which flight is no longer possible. They based this conclusion upon a study of steamer ducks, a group in which three species are flightless and one, *Tachyeres patachonicus,* flies badly with great difficulty at takeoff. The flightless species have very high wing loadings (averaging from 2.6 to 4.87 g/cm^2) whereas the only steamer duck that can fly has an average wing loading between 1.87 and 2.22 g/cm^2.

Another important predictor of aerial performance is the aspect ratio of the wing, measured as the length of the total wingspan (including the trunk segment between the two wings) divided by the average breadth of the wing. In birds, as in aircraft, wings with a short span that are very tapered—and thus on average rather narrow—have a low aspect ratio, in the range of about 4 or 5. This wing shape is associated with fast flight and great maneuverability, and can be found in birds like the peregrine falcon, which preys on other birds, usually catching them in the air. Long wings with high aspect ratios are typical of long-range, relatively slow-cruising aircraft, such as gliders that may have aspect ratios of 12 to 15, or soaring birds like condors, which have an aspect ratio of about 20 (Figure 71). Large flying organisms, according to Pennycuick, have few options. Although the largest living birds that can fly come from different avian families, including the storks, vultures, bustards, and pelicans, they all converge on a similar airfoil shape: their wings are long and relatively narrow, with the primaries spread apart at the tips like fingers on an opened hand, and they have a very high aspect ratio. Longer wings are better for sustained, horizontal flight—especially soaring—because a long, narrow wing produces more lift relative to drag than do short or broad wings. Thus birds with high

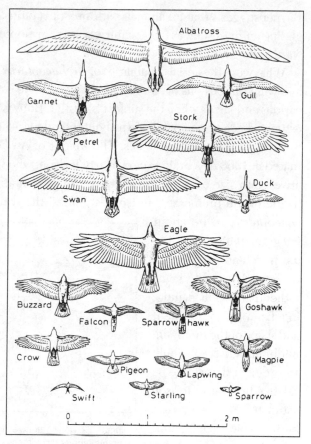

Figure 71. One determinant of a bird's aerial performance is its aspect ratio. Birds with high aspect ratios, like albatrosses, specialize in long-distance soaring flight; birds with short tapered wings have low aspect ratios that emphasize maneuverability, like the fast-flying, predatory falcon. Broad, short wings, like those of the eagle or the goshawk, provide rapid takeoff and good maneuverability.

aspect ratios can soar at shallow angles, which means they travel farther horizontally for every foot they drop in height. Longer wings also power faster flight, while compromising the bird's ability to make tight turns and take off from level surfaces. In contrast, short, broad wings can be flapped more rapidly for better takeoff and provide more maneuverability, which makes them useful for birds that fly among vegetation or trees. If condors and similar birds represent the maximum aspect ratio, there is also a minimum. Jeremy Rayner speculates that level flight may be impossible for vertebrates with aspect ratios lower than about 3 or 4, because the wings will produce more drag than lift.

Rayner suggests that for animals with an aspect ratio of about 1.5—like typical gliding animals such as flying squirrels and dermopterans—only slow, downward gliding is possible.

What were the aerial capabilities of *Archaeopteryx*? To find out, the weight and wing dimensions of *Archaeopteryx* have to be estimated: somewhat precarious endeavors when the subject is extinct. Body weight is relatively easy to estimate because much is known about the relationship of body size to various linear dimensions of the bones. The weight of *Archaeopteryx* was first estimated in 1968 by a brain expert, Harry Jerison, who was interested in the relative brain size of *Archaeopteryx*. Working from an endocast (a cast of the inside of the braincase), Jerison estimated the weight of the brain of *Archaeopteryx* to be about 0.9 gram (0.03 ounce) and its shape and proportions to be "obviously avian" rather than reptilian, judging from the placement and shape of the forebrain, midbrain, and cerebellum. In order to determine its relative brain size, Jerison estimated *Archaeopteryx*'s body weight as 500 grams (about a pound) "from reconstructions" without specifying exactly how he proceeded. This value

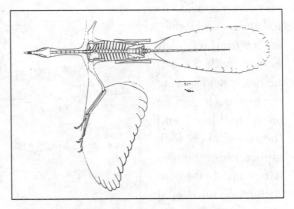

Figure 72. Derik Yalden estimated the weight (about 250 grams) and bodily dimensions of *Archaeopteryx*, based primarily on the Berlin specimen. With a wingspan of 58 centimeters, an aspect ratio of 7, and a wing area of 479 square centimeters, Yalden's Archaeopteryx was small-winged: a poor flier with difficulty taking off.

made the brain of *Archaeopteryx* relatively larger than in modern reptiles but not yet quite as large as in modern birds. Jerison concluded that *Archaeopteryx* had thus evolved an avian organization to the brain but had not yet completed the expansion in relative brain size that would make it fully avian. (Later and probably more reliable estimates of body weight are smaller, about 200–250 grams [7–8.75 ounces], which gives *Archaeopteryx* a larger ratio of brain size to body size that falls on the edge of those recorded for modern birds.)

The body weight of *Archaeopteryx* was reestimated by Derik Yalden in 1971, and then revised again by him in 1984 (Figure 72). Yalden's second estimate is the most rigorous and systematic, and has been widely accepted. Starting with the skeletal dimensions of the Berlin specimen, Yalden estimates body size by plugging *Archaeopteryx's* measurements into mathematical formulae that describe the relationships among various bony dimensions and body weight in a wide range of living birds and bipedal mammals. (Bipedal mammals were included because the bones of *Archaeopteryx* may not have been as fully pneumatized as bird bones; *Archaeopteryx* may have been more comparable to a mammal than a bird in this regard.) In all, Yalden derives twenty-six estimates of the body weight of *Archaeopteryx*, finding the most reasonable to be about 250 grams. This value falls close to the middle of the range for either avian or mammalian models. Many later workers have accepted either 250 or 200 grams as the weight of *Archaeopteryx*, following Yalden.

Using this body weight, Yalden estimates the following dimensions and aerodynamic properties for *Archaeopteryx*:

Table 1. Yalden's Dimensions and
Aerodynamic Parameters of Archaeopteryx

Body weight	250 g
Wingspan	58 cm
Individual wing length	23.5 cm
Individual wing breadth	8.25 cm
Two wing area	388 cm²
Total wing area	479 cm² (includes intervening body strip)
Tail area	159 cm²
Wing loading: two wing area	0.64 g/cm²
Wing loading: total wing area	0.52 g/cm²
Aspect ratio	7.0
Stalling speed°	5.1–6.2 meters/second (midpoint = 5.65 m/sec)

° Calculated by Padian, not Yalden.

The numbers sketch the proportions and dimensions of *Archaeopteryx* fairly fully. Because Crawford Greenewalt, writing in 1962, showed that the proportions of living birds are remarkably consistent and mathematically predictable, the proportions of Yalden's *Archaeopteryx* can be compared to those of various groups of living birds, but I found that no one had yet done this research. I wanted to answer two questions: first, does Yalden's *Archaeopteryx* fall within the known boundaries of living birds, and second, if so, what sort of bird does *Archaeopteryx* resemble?

Greenewalt's most reliable formulae are based on a grouping of birds he calls "passeriforms," which includes a wide range of birds from twelve different avian families, including herons, falcons, hawks, vultures, owls, and eagles as well as typical perching birds like finches, wrens, thrushes, starlings, crows, and so on. He has a second grouping that he calls "shorebirds," which includes doves, parrots, geese, swans, two bustards, a kingfisher, and an albatross as well as classic shorebirds like plovers, oystercatchers, curlews, and sandpipers. This second group apparently has different characteristic wing proportions, although Greenewalt cautions that the data on which the shorebird formulae are based are much less numerous, making his equations less reliable. Finally, Greenewalt's third grouping is "ducks," which includes grebes, loons, and coots in addition to various species of duck. The duck model describes yet a third typical shape for birds, but the duck model is based on the smallest amount of data.

Using Greenewalt's formulae, I predicted the size and shape of the wings and values for aerodynamic properties of *Archaeopteryx* if it were shaped like a modern bird of each group. I then compared these "normal" bird values to those in Yalden's reconstruction of *Archaeopteryx*. The results are shown in Table 2.

Table 2. Archaeopteryx *Compared to Three Groups of Living Birds*

	Archaeopteryx	Passeriform	Shorebird	Duck
Wing area (cm²)	388 (two wing) 479 (total)	696	478	241
Wing loading (g/cm²)	0.64–0.52	0.28–0.29	0.52	1.04
Aspect ratio	7.0	7.2	5.7	5.7
Wingspan	58 cm	59 cm	69 cm	69 cm

Compared to the passeriform model, Yalden's reconstruction has a very low wing area that falls between 55 percent and 69 percent of the predicted wing size. Such a wing area would be typical of a bird that weighed only 119–155 grams (4.2–5.4 ounces): a much smaller body weight than that estimated for *Archaeopteryx*. And because Yalden's *Archaeopteryx* is distinctly small-winged for its body size, it also has a very high wing loading (129 percent of the predicted value). However, Yalden's *Archaeopteryx* has almost exactly the predicted wingspan. The discrepancy in estimated and predicted wing size shows how odd the proportions of this reconstruction of *Archaeopteryx* are for this model.

In contrast, Yalden's *Archaeopteryx* fits the shorebird model perfectly in terms of the proportions of weight, wing size, and wing loading, although its aspect ratio is high and its wingspan is shorter than predicted. Yalden's *Archaeopteryx* looks a great deal like a "shorebird" (under this loose definition of shorebird). This finding is particularly interesting because Richard Thulborn and Tim Hamley of the University of Queensland actually suggested that *Archaeopteryx* might have had an ecological niche similar to extant shorebirds (Figure 73). They wrote:

> *Archaeopteryx* is envisaged as an agile hunter that frequented shore-lines, pools, and shallow waters. It probably subsisted on prey of moderate size, including small fishes and worms. The feathered forelimbs ("wings") may have been used as a canopy while *Archaeopteryx* foraged in water; its hunting techniques may have resembled those used by existing herons and egrets. . . . the first flying birds may have been aquatic forms, using their rudimentary flight apparatus to carry them from wave-crest to wave-crest.

Although their suggestion has not been widely embraced, Thulborn and Hamley may have realized intuitively that Yalden's *Archaeopteryx* has shorebird-like proportions. However, *Archaeopteryx* has wings that are narrow for their span (giving an aspect ratio of 7.2 versus a predicted 5.7) compared to shorebirds or ducks with comparable wing areas or wingspans.

Figure 73. Paleontologists Richard Thulborn and Tim Hamley suggested that *Archaeopteryx* might have been a wading predator that used its wings as a canopy over the water while hunting.

Relative to the duck model, Yalden's *Archaeopteryx* is very large-winged (160–200 percent the predicted area); it has wing loading only about half that predicted, a short wingspan, and a high aspect ratio, which can again be attributed to great narrowness of the wings. Typically, the birds included in the duck model are large-bodied, small-winged, and have a high wing loading; they are fast, enduring fliers but are poor at takeoff. *Archaeopteryx* does not match this physical description at all well and probably did not share these aerial abilities. The misfit between the proportions of *Archaeopteryx* and the duck model becomes clear when weight is calculated backwards from the wing area. With wings as large as Yalden has estimated, a "duck" would weigh between 653 and 4,643 grams (23 ounces and 163 pounds)—much, much more than *Archaeopteryx* can possibly have weighed.

In summary, Yalden's *Archaeopteryx* makes a strange "passeriform," a good but long-winged "shorebird," and a "duck" so odd that the possibility can be completely dismissed. What implications about *Archaeopteryx*'s abilities can be derived from these comparisons? By "passeriform" standards—and most of the birds I think of as ordinary birds fall into this category—Yalden's reconstruction is small-winged and has high wing loading. Both small wing size and high wing loading directly limit a bird's ability to create lift, making takeoff and sustained flight difficult. Thus, relative to the birds included in Greenewalt's passeriform model, Yalden's *Archaeopteryx* is a poor flier indeed. But relative to the birds in the shorebird model, Yalden's *Archaeopteryx* is almost average, because the birds in this model are not particularly adept fliers. Only in aspect ratio and wingspan does Yalden's *Archaeopteryx* match the passeriform model better than the shorebird model. Among shorebirds, a higher-than-average aspect ratio and short wings are seen in birds like plovers, godwits, swans, and geese, fast fliers that often have some difficulty in takeoff. Similar capabilities can be postulated for Yalden's *Archaeopteryx*.

Another indicator of *Archaeopteryx*'s aerial abilities is stalling speed, which can be regarded as the minimum speed a "ground-up" flier must attain in order to achieve takeoff and then gain altitude. Stalling speed is the velocity at which lift starts to diminish rapidly. Stalling occurs when the airflow separates from the surface of the airfoil and lift starts to decline; when lift is no longer greater than the weight of the animal, then the animal begins to fall. Clearly, a flying animal must achieve and maintain a velocity greater than the stalling speed if it is to fly, the sole exception being in landing when a controlled stall is often useful. Stalling speed is a function of body weight and wing area (or wing loading), the density of air, and maximum lift. Using Yalden's reconstruction of *Archaeopteryx*, Padian estimated its stalling speed at 5.1–6.2 meters per second (m/sec). Since a small, bipedal, terrestrial animal can run at about 3–4 m/sec, a takeoff run would leave between 1 and 3 m/sec to be generated by flapping. With the short and small wings of Yalden's reconstruction, it looks as if *Archaeopteryx* may have faced real difficulty in generating its minimum stalling speed and taking off from the ground.

While wing area or airfoil size is obviously an important determinant of the ability to take off, another crucial variable—and the one that Yalden refrains from estimating numerically—is the size of its flight muscles. This value is not readily derived from planar dimensions. Since 1954, when Gavin de Beer made the argument in his monograph on the London specimen, many have believed that *Archaeopteryx* could not fly. The evidence was its lack of a bony, keeled sternum (revealed to exist in the most recently found specimen); the small area for the insertion of the pectoral muscles on the humerus; and the short coracoids, which also anchor flight muscles. All of these features imply that the flight muscles on *Archaeopteryx* were deficient. Yalden takes issue with these assessments, observing that the coracoids are indeed short but that the insertion on the humerus is both very wide and very long compared to a sample of modern birds. To him, this suggests a short and broad pectoral musculature equal in power-producing ability to that in modern birds. A similar conclusion has been reached by Storrs Olson and Alan Feduccia, writing in 1979, who suggested that the unusually robust furcula of *Archaeopteryx* served as extra bony attachment for a strong pectoralis muscle, which compensated for the shortness of the coracoids. Later observers also pointed to the lack of a triossial foramen (the structure in the shoulder that enables a bird to perform a wing flip) and the accompanying modification of the supracoracoideus muscle (which also powers the upstroke) as evidence that flight was impossible. However, Ostrom counterargued that the upstroke could be powered by other muscles (primarily the deltoideus) in the shoulder region; his analysis, and greater acquaintance with Sy's experiments in severing the tendon of the supracoracoideus of pigeons (who were nonetheless able to sustain horizontal flight), convinced most researchers that *Archaeopteryx* was anatomically capable of flight, if not necessarily takeoff. Yalden takes this view also, concluding that *Archaeopteryx* may have had flight muscles comparable in size and power to those of living birds, without estimating their size precisely. Even with powerful but small flight muscles, *Archaeopteryx*'s small wings and high wing loading make it a poor candidate for takeoff from the ground. Yalden's assessment is that it needed an arboreal perch for takeoff, after which it could perhaps glide or flap clumsily only over short distances.

This inexact assessment of *Archaeopteryx*'s aerial abilities held sway as the many debates over *Archaeopteryx* came to a climax in 1984, when an international conference was convened in Eichstätt, Germany. All of the main researchers presented their points of view to one another and argued ardently. Fittingly, a special participant at the conference was the Teyler specimen of *Archaeopteryx*—the one that John Ostrom had recognized so suddenly in 1970—which was carefully hand-carried to the meeting by its curator, John de Vos. The Eichstätt meeting closed with an earnest attempt to reach a synthesis, which was summarized later: "*Archaeopteryx* was an active, cursorial predator and was also facultatively arboreal; it was a glider and a feebly powered, or flapping flier. Finally, it was incapable of takeoff and flight from the ground upward. . . ." Obviously, this scenario contains many contradictory elements and unresolved disputes. Most participants were persuaded by *Archaeopteryx's* hindlimb morphology and anatomical resemblances to theropod dinosaurs that *Archaeopteryx* was a bipedal cursor, but they were also persuaded by the modern feathers, forelimb structure, and avian wings that it was a flapping flier. However, given the lack of an avian shoulder, the indications of small flight muscles, and the lack of a keeled sternum (which was not yet known to exist), takeoff from the ground seemed impossible. As one observer concluded, "[T]he concept of flight from the ground up was left in shreds." Because

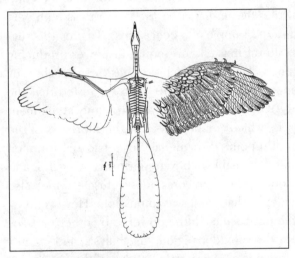

Figure 74. Siegfried Rietschel's new estimate of *Archaeopteryx's* wing area (with feathers indicated) is almost 60 percent larger than Yalden's (solid outline). If Rietschel is correct, then *Archaeopteryx* was a more maneuverable flier with a better ability to take off and a low stalling speed.

the legs said "ground" and the wings said "trees," the conferees tried to imagine an animal with legs suited for terrestrial bipedalism that had to climb trees in order to fly. It was an awkward compromise. They tried to weld their beliefs into a consensus and instead created an unviable chimera.

What went largely unnoticed at that conference were the implications that could be drawn from Siegfried Rietschel's presentations. While the linear dimensions of Yalden's *Archaeopteryx* were based on the bones—and therefore undoubtedly correct—the dimensions of the wings were more open to interpretation. Rietschel's main paper at the Eichstätt conference was a close look at the wings and feathers of *Archaeopteryx,* not for aerial abilities but with an eye to the exact arrangement of the feathers; in a shorter paper, Rietschel examined the invidious charges of feather forgery. Rietschel's work is probably the most painstaking examination of the wing and feathers of the Berlin *Archaeopteryx* ever carried out; it is detailed, thorough, meticulous, and difficult to challenge. In it, Rietschel draws a new reconstruction of the right wing of *Archaeopteryx,* showing each feather on the wing, extended as if in flight (Figure 74). But this reconstructed wing is much broader—about 57 percent broader—than Yalden had estimated, although few of the conferees realized this. This expansion of the wing in turn increased the wing area, from Yalden's estimate of 388–479 cm^2 (depending upon whether the intervening body strip is excluded or included) to 611–754 cm^2.

Changing the dimensions of the wing so dramatically obviously altered the aerodynamic parameters of *Archaeopteryx,* too. In Table 3, I give the new dimensions and calculate the aerodynamic parameters of *Archaeopteryx* based on Rietschel's work for comparison with Table 1, based on Yalden's reconstruction.

Table 3. Dimensions and Aerodynamic Parameters of Archaeopteryx *from Rietschel's Reconstruction*

Parameter	Rietschel's Estimate	Difference from Yalden's Estimate
Body weight	250 g	0%
Wingspan	58 cm	0%
Individual wing length	23.5 cm	0%
Individual wing breadth	13.0 cm	+58%
Two wing area	611 cm^2	+57%
Total wing area	754 cm^2	+57%
Wing loading: two wing area	0.41 g/cm^2	-36%
Wing loading: total wing area	0.33 g/cm^2	-37%
Aspect ratio	4.5	-36%
Stalling speed	4.1–5.0 m/sec	-20% (approx.)

With broader wings and thus a larger wing area, *Archaeopteryx* is a different-looking sort of bird with markedly different abilities. Whereas Yalden's reconstruction is a small-winged bird with a heavy wing loading, similar in proportions to Greenewalt's "shorebird" model, the Rietschel *Archaeopteryx* is close to the "passeriform" model, which predicts a wing area of 696 cm² and a wing loading of 0.33 g/cm². In both of these parameters, the takeoff abilities of Rietschel's *Archaeopteryx* are improved over Yalden's reconstruction. With its much broader wings, the new reconstruction of *Archaeopteryx* has a much lower aspect ratio: 4.5 versus the predicted value of 7.0. A low aspect ratio diminishes flight speed and makes soaring or gliding a less feasible endeavor, both changes that increase the energetic expense of flight. In this regard the new reconstruction of *Archaeopteryx* is a less efficient and slower flier than is Yalden's reconstruction. However, on the positive side, a low aspect ratio (like large wing area and low wing loading) facilitates takeoff and increases maneuverability, a useful compromise for birds that inhabit vegetated areas. Combined with a low wing loading, these traits typify birds that hunt and may need to carry their prey.

Finally, this new reconstruction and Yalden's differ in terms of their estimated stalling speeds. Stalling speed for the new reconstruction is about 4–5 m/sec, roughly a 20 percent decrease from the 5–6 m/sec needed by

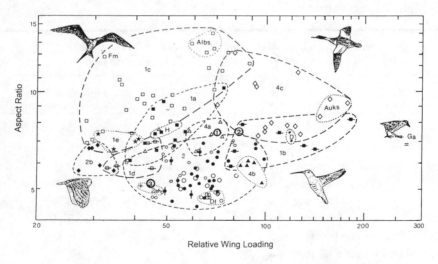

Figure 75. Living birds with similar foraging habits have similar values for their relative wing loadings and aspect ratios. ① A 200-gram *Archaeopteryx* with Yalden's estimated proportions falls within a cluster of small-winged shorebirds that forage in open spaces, or at the edge of a group of birds including woodpeckers, titmice, wrens, and corvids that forage in closed habitats. ② A 250-gram *Archaeopteryx* with Yalden's estimated proportions falls on the edge of the shorebird group and another cluster that includes swans, geese, ducks, and loons. ③ A 250-gram *Archaeopteryx* with Rietschel's larger wing size falls within a different cluster of birds—ones that fly within vegetation—and is particularly close to flycatchers.

Yalden's *Archaeopteryx*. The new reconstruction also has much larger wings with which to produce lift, which increases the probability that *Archaeopteryx* could generate enough lift for takeoff even if its flight muscles were small. *Archaeopteryx* would have to generate only 0–2 m/sec over running speed in order to take off from the ground, which seems highly feasible given its large wings.

There is another issue that is highlighted by the differences between Yalden's *Archaeopteryx* and this new reconstruction. The Norbergs, a husband-and-wife team in Göteberg, Sweden, have studied various aspects of flight for years. Using data on many different living birds, they have shown that a combination of aspect ratio and wing loading will cluster species into groups united by the flight style and foraging mode of their members. These clusters are not based on phylogenetic relationships but on the habits and adaptations of the species in question (Figure 75). Because wing loading scales with body size—larger species have higher wing loading—the Norbergs correct the wing loading for body size in their work. Using Yalden's basic dimensions for a 200-gram *Archaeopteryx*, with wings of 479 cm² and an aspect ratio of 7, Ulla Norberg calculates a relative wing loading of 71.

These values place *Archaeopteryx* in a fairly central position in a tight cluster of birds with a flight style described as "ground foraging in open spaces" or at the very edge of a diffuse cluster characterized as flying "among vegetation." Norberg apparently feels that the latter group is more likely and discusses *Archaeopteryx* as being a bird with "somewhat higher aspect ratio and relative wing loading than average values for arboreal birds . . . [but] not exceptional [for this category]. . . ." Birds that fit this closed-habitat cluster include woodpeckers, nuthatches, creepers, titmice, wrens, wagtails, starlings, corvids, and thrushes. According to Norberg, the members of this group have broad, short wings and a low aspect ratio that give them slow, highly maneuverable, and expensive flight. Unfortunately, this does not sound like an accurate description of Yalden's *Archaeopteryx* and its aerodynamic parameters. However, the values for Yalden's *Archaeopteryx* are actually closer to the alternative grouping, the cluster of species that forage in open spaces. This group includes sandpipers, plovers, and similar long-legged, small-winged shorebirds, which are characterized as having fast and relatively cheap flight, suitable for commuting or migration. Their wing characteristics are: "above-average aspect ratio, rather high wing loading, rather short wings," which describes Yalden's *Archaeopteryx* well.

Replotting *Archaeopteryx* on Norberg's graph using a 250-gram body weight and Yalden's reconstructed wings puts *Archaeopteryx* in a slightly different location. The point representing its proportions falls on the edges of both the shorebird group and another similar cluster, characterized as "foraging in open water," which includes swans, geese, ducks, and loons. What is interesting is that Yalden's reconstruction consistently falls among or close to birds that live near water—and the Solnhofen sediments were, after all, deposited in a lagoon. The grouping is also supported by the body and wing proportions of Yalden's reconstruction, which approximate closely the proportions in Greenewalt's "shorebird" model. But what difference does the larger wing size of Rietschel's *Archaeopteryx* make? Using the values obtained from his reconstruction, I replotted *Archaeopteryx* once again. *Archaeopteryx* now moves to a more central location within the cluster of birds that fly within vegetation; the point for *Archaeopteryx* falls in or very near to a subcluster consisting of the flycatchers. Norberg describes the habits of flycatchers as "perching and hawking among vegetation" with "slow, highly maneuverable, but rather expensive flight"; physically, they have a "low aspect ratio, short wings, low wing loading." This description neatly matches the abilities deduced for Rietschel's *Archaeopteryx* as well.

I realized there were some important aerodynamic implications of Rietschel's reconstruction of *Archaeopteryx*. First, *Archaeopteryx* was built like birds that fly in closed, heavily vegetated habitats and capture live prey, such as insects or small animals. Being predatory is consistent with an ancestry among the maniraptors or theropods; it suggests that *Archaeopteryx* (and the early part of the avian lineage) may have retained the same ecological niche while evolving a new way to capture prey. Thus both the conclusion that *Archaeopteryx* could take off from the ground and its inferred ecological niche offer support to the "ground up" hypothesis. However, most of the living birds that use closed habitats and fly among vegetation are arboreal, frequently climbing trees (like woodpeckers and their allies) or perching (like flycatchers)—a fact that offers support to the "trees down" hypothesis. This finding may explain the contradictory elements in the consensus view of *Archaeopteryx* at the close of the Eichstätt conference. The truth may indeed be that *Archaeopteryx* had features of both ground-dwelling dinosaurs and arboreal birds, perhaps because it was in transition between the two.

Second, *Archaeopteryx* in this reconstruction was a slow flier that incurred a high metabolic cost for its flights. While fast flight is easier in the sense of being less expensive and thus more efficient, it seems improbable that a lineage's first evolutionary forays into a new mode of locomotion will be maximally efficient. In any form of locomotion, the ability to attain high speeds is generally regarded as a late specialization, not the entry route. For example, fast-running cheetahs are regarded as more specialized and evolutionarily advanced in their locomotor system than the slower-moving lions. Following this principle, early fliers may well have been capable of only slow and expensive flight, evolving the anatomical apparatus for energy-saving (like wing flips and high aspect ratios) later.

Third, if this was indeed *Archaeopteryx*'s habitat and style of foraging, it was using an ecological niche that was aerial but was nonetheless distinctly different from the niche already occupied by pterosaurs. Generally, pterosaurs are seen as soaring over open water and spearing fish near the surface; their aerodynamic adaptations are for long, low-cost soaring flights in open spaces. If pterosaurs were already filling that ecological niche, it makes a satisfying kind of sense that the earliest birds would be forced to specialize in an aerial niche that would minimize competition from pterosaurs. Flying among shrubs, bushes, or low plants, slowly and expensively and for short distances, can be seen as the ecological opposite of the open habitat, soaring, and long-distance pterosaur niche. Since pterosaurs could not easily fly among vegetation or maneuver quickly after prey, this niche would have been open for exploitation by another sort of flying animal.

But a stumbling block remained. It would be easy at this point to simply conclude that *Archaeopteryx* could and did fly like the woodland birds whose proportions it mimicked so closely. But the small pectoral muscles and the lack of an ability to perform a wing flip were troubling. Maybe it could fly, but could *Archaeopteryx* take off? Did those modern feathers and large wings compensate for the skimpy flight muscles? Or was *Archaeopteryx* restricted to takeoff from a raised perch?

Chapter 12.
The Tangled Wing

———

The problem was how to answer these highly specific questions with any certainty. Then, a few years after the Eichstätt conference, Jim Marden recognized the extreme significance of takeoff to a flying adaptation. Takeoff is obviously one of the most important phases in flight, for all flight begins with takeoff—and no flight can continue if takeoff fails. Takeoff is also perhaps the most physically demanding aspect of flight. So how could he find out which of the variables involved in aerial performance actually affects takeoff most directly?

Marden took an empirical approach to the problem, collecting and analyzing data on living insects, birds, and bats. Marden wanted to focus on an attribute that could be measured rather than subjectively assessed. Watching birds take off in the wild and deciding what was good, sufficient, or poor would be too arbitrary. His insight was that an animal's ability to take off from a standstill in still air is a function of the power and lift the animal can generate by flapping its wings. He called the net upward force an animal can exert upon its own body by flapping its wings *maximum lift*. (Lift in this sense applies to the entire body and not just the airfoil.) When maximum lift is more than the animal's body weight, then the animal can take off from the ground or anywhere else. If the maximum lift is less than the animal's body weight, then it cannot take off from level ground and must be assisted by the force of gravity.

To find out what maximum lift was, Marden attached lead weights to the abdomens of insects, the thighs of birds, and the lower backs of bats—147 species in all—and then encouraged them to try to take off. He started with a weight that added 20 percent to the animal's body weight and continued to add others of the same size until the animal could no longer take off. Each animal had three attempts at takeoff at each weight and could rest between attempts. Sometimes animals jumped upward when they clearly could not achieve a proper takeoff, so he defined a successful takeoff as one producing aerial locomotion over a distance greater than two body lengths. For each animal, Marden recorded a series of variables. The first was maximum mass, defined as halfway between the largest mass the animal could get off the ground (body weight plus lead weights) and the mass produced by adding one additional weight, which made takeoff impossible. Marden then calculated the force needed to achieve takeoff at maximum mass: maximum lift. He also measured the angle of takeoff since a vertical takeoff indicates high proficiency, a low-angle takeoff indicates greater difficulty, and a takeoff at an angle of zero degrees is simply forward motion. He recorded total body weight, wing area, wingspan, aspect ratio, and the ratio of the flight muscles to the total body weight (including the lead weights) for each experiment.

Analysis of these data yielded unusually clear conclusions, for the ability to take off is almost completely dependent upon the flight-muscle ratio. Few of the other variables that Marden considered have a consistent or significant effect. Among the birds and bats, the best takeoff performance occurred in animals with flight-muscle ratios between 0.20 and 0.35 (or 20–35 percent of their total laden body weight), while the best-performing insects had flight-muscle ratios up to 0.56. Animals with such high flight-muscle ratios achieved vertical takeoffs and could lift up to three times their own body weight. Fliers with flight-muscle ratios of less than 0.20 achieved takeoff only at low angles and could lift only small additional weights. When the flight-muscle ratio fell to between 0.16 and 0.18, takeoff became marginal. An unanticipated observation was that animals burdened with loads so heavy that they could barely take off were still able to fly around more or less indefinitely once they were airborne, showing how much more difficult it is to take off than to sustain flight once airborne. Thus Marden suggests that the takeoff problems experienced by steamer ducks may be due to their exceptionally low flight-muscle ratios—0.153 to 0.173 for the flightless species and 0.201 for the flying steamer duck—rather than to their high wing loading. Only *Tachyeres patachonicus* has a high enough flight-muscle ratio to take off. But Livezey and Humphrey are probably correct that, once airborne, this species' high wing loading greatly hampers its aerial performance. Because its wings are so small, *Tachyeres* must expend extra energy flapping, making its flight very expensive.

Marden treats the marginal flight-muscle ratio of 0.16 as a Rubicon, below which flight becomes extremely difficult and ultimately impossible. This notion is supported by the findings of other researchers, too. For example, Pennycuick conducted a study of the Kori bustard, a large African bird weighing about 12 kilograms (26.5 pounds). Kori bustards are notoriously poor and awkward fliers. They need a long taxiing run to achieve takeoff, and even once airborne, they fly with apparent difficulty for only short distances. As Marden's work predicts, Kori bustards have a flight-muscle ratio of 0.164, very close to the ratio at which takeoff becomes impossible. Another indication of the reality of this threshold comes from data on many living bird species. Out of 425 species with reported flight-muscle ratios, only twelve (3 percent) have flight-muscle ratios lower than 0.16. Some of these species with low flight-muscle ratios are aquatic or semiaquatic species known to be poor fliers, like the least grebe or the American coot; in these cases, the low flight-muscle ratio is correlated with poor flight abilities. But some of the other birds are reclusive, inhabiting dense foliage, so that there are few field observations of their flight capabilities. As Marden puts it, "These birds should be of special interest for future studies, to determine if and how they derive the necessary amount of lift from their relatively small flight musculature." Unfortunately, there was no obvious way to apply these results to the paleontological dilemma of *Archaeopteryx* at the time, because no one knew the flight-muscle ratio of *Archaeopteryx*.

The next step forward in assessing the aerial abilities of *Archaeopteryx* came in 1991, with a startling paper published by John Ruben in one of his first forays into the intersection of physiology and paleontology. Ruben was struck by the inherent contradictions in the consensus scenario from Eichstätt, which he attributes to the underlying assumption that *Archaeopteryx* was warm-blooded and endothermic, like later birds. His novel suggestion is that perhaps *Archaeopteryx* was cold-blooded—not cold-blooded like the passive and sluggish reptile of stereotype, but cold-blooded as many modern reptiles truly are. As Ruben correctly observes, since *Archaeopteryx* is morphologically very reptilian, with toothy jaws, a reptilian shoulder girdle, and a long, bony tail, it may well also have been reptilian in physiology.

The two primary pieces of evidence that *Archaeopteryx* was warm-blooded are its feathers and its expensive mode of locomotion (whether it is a cursorial, terrestrial biped or a fully flapping flier). These facts seem to point to endothermy and must be discounted or explained away if *Archaeopteryx* is to be considered cold-blooded. Ruben addresses each issue in turn, first arguing that feathers or other body covering is not essential for regulating body temperature, as the effective basking habits of reptiles show. Feathers do not prevent all heat gain nor do they ensure heat retention.

> Consequently [argues Ruben], feathers in *Archaeopteryx,* or any other bird, do not necessarily imply a fixed capacity for either deflection or retention of internally or externally generated heat. A fully feathered *Archaeopteryx* could have been either ectothermic or endothermic.

It is important to appreciate Ruben's point exactly. He does *not* suggest that any known feathered species is cold-blooded or that *Archaeopteryx* cannot have been warm-blooded, like every other known feathered creature. What he suggests is that the one-to-one equivalence of feathers and endothermy is not obligatory: that a feathered animal *can* be cold-blooded. To bolster this interpretation, he emphasizes the fact that *Archaeopteryx* may not have an enhanced respiratory system with air sacs within its hollow bones, a feature that, when present, is closely linked to the high rates of lung ventilation needed to fuel high rates of aerobic metabolism. (An alternative possibility, not discussed by Ruben, is that the features of modern avian endothermy evolved as a mosaic, with the respiratory enhancements and high metabolic rates coming after basic insulatory mechanisms had evolved to help control body temperature.)

Ruben makes an argument of similar subtlety about *Archaeopteryx*'s lifestyle and mode of locomotion, which he believes is based on a false and limited view of reptilian physiology and behavior. He argues that statements that reptiles are sluggish or can maintain bursts of rapid locomotion for only seconds are simply inaccurate. He explains,

> Certainly, compared to modern birds and mammals, reptiles generally rely more heavily on fatigue-inducing anaerobic metabolism during periods of intense exercise. . . . However, numerous species . . . have expanded metabolic capacities, maintain high levels of diurnal activity and alertness, patrol large home ranges, and rely on their speed and endurance to actively pursue prey. . . . Additionally, a wide variety of reptiles . . . are capable of main-

taining burst levels of exercise, *ranging to several minutes in du-*
ration. . . . Moreover, during the initial phases of burst activity,
the metabolic power output and locomotory speed of some reptiles exceeds that of equal-size mammals. . . .

Not only could *Archaeopteryx* have had a reptilian physiology, according to
Ruben, but reptilian physiology is also much less limiting than is widely
supposed.

The surprising advantage of reptilian physiology, according to Ruben, is
that in "burst-level" activity—that is, anaerobic activity that produces a short
burst of energy, such as takeoff—some reptiles can generate at least twice as
much power per gram of muscle than is generated by birds and mammals.
The obvious implication is that a reptilian *Archaeopteryx* with flight muscles
half the size of a similar-sized bird could produce just as much energy for
takeoff as that bird does. Thus small flight muscles would not necessarily prohibit a reptilian *Archaeopteryx* from flying—and Ruben knew that there was
evidence that *Archaeopteryx*'s pectoral muscles were small by bird standards.

To explore exactly how much power *Archaeopteryx* might be able to generate—and might need to fly—Ruben arbitrarily set the flight muscles at 7
percent of body weight. He chose this value because it represents "a little less
than half" the minimum flight-muscle ratio in most modern birds. He selected
this flight-muscle ratio of 0.07 as a heuristic device, not as a realistic estimate,
because he wanted to show how small the flight muscles of *Archaeopteryx*
could be and still be effective. Ruben also allowed an additional 2 percent for
shoulder muscles, which (due to *Archaeopteryx*'s unusual anatomy) probably
performed a more crucial role in flight than they do in modern birds. A flight-
muscle ratio of 0.07 to 0.09 is very low; in fact, it is well below the bottom of
the range observed in modern birds, so this figure places Ruben's *Archaeopteryx* entirely outside the known limits of avian anatomy. Still, with a
reptilian physiology, an *Archaeopteryx* with a flight-muscle ratio of 0.07–0.09
would produce as much power for takeoff as a bird with a flight-muscle ratio
of about 0.14–0.18. The implication is that *Archaeopteryx* would have been as
adept at takeoff as birds like the Kori bustard or the steamer duck, who may
struggle but nonetheless get off the ground.

Was sustained flight possible, too? To address this question, Ruben decided
to look at the minimum power velocity for an *Archaeopteryx* weighing 200
grams (7 ounces). As explicated by Pennycuick, the energy expended in flying
can be diagrammed as a curve. At low velocities—such as stalling speed, the

minimum velocity above which horizontal flight is sustained—the energy demands may be high; similarly, at high velocities, energy is expended at high rates. But somewhere between the two lies the minimum power velocity, at which flight is cheapest. Ruben calculated that 7.7 meters per second (or 8.2 m/sec for a 250-gram *Archaeopteryx*) would be the minimum power velocity for *Archaeopteryx*. He reasoned that if *Archaeopteryx* could not muster enough energy to fly at the minimum power velocity, it certainly couldn't fly any faster—or slower, either of which would require more energy. Ruben calculated that a 200-gram *Archaeopteryx* could generate roughly twice as much power as it would need to achieve minimum power velocity, *if* it produced power at its maximum output, *if* it had a flight-muscle ratio of at least 0.07, and *if* it had a reptilian physiology. This finding was strong support for the idea that a reptilian *Archaeopteryx* could take off and fly handily.

But there is still another problem to be faced: working at maximum power output generates significant fatigue, and reptiles and other ectotherms recover slowly from such fatigue. Was *Archaeopteryx* constantly ending its flights out of control because the metabolic demands of sustaining flight were so great? Ruben estimated that a maximum flight for *Archaeopteryx* might have been as much as 1.5 kilometers (nearly a mile)—although, he warned, such a prolonged flight would induce severe postflight fatigue and require a recovery time of one hour or more (if *Archaeopteryx* had recovery rates approximately equal to those of modern reptiles). He believes such long flights were probably rarely, if ever, performed. Short flights, on the order of 10–20 meters (roughly 33–40 feet), would have required only about sixty seconds of recovery time and seem more reasonable to Ruben. At a velocity of 7.7 or 8.2 m/sec, a 20-meter flight would take roughly one to two and a half seconds.

I read these estimates and wondered about their consequences in real life. To me, a minute of recovery time after a two- or three-second flight seems like a serious problem, especially if the end of the flight—the point at which fatigue is greatest—involves a landing that must be carefully controlled. Ruben's recovery rates are necessarily based on terrestrial, not aerial, animals; he has no other choice. But it is obviously true that the consequences of stopping running because of fatigue are far less severe than those of stopping flying abruptly. To me, it seemed likely that Ruben's reptilian *Archaeopteryx* could fly only over very short distances in order to save enough energy for landing as well. While flight of less than thirty-three feet is not impressive, it is flight nonetheless. If Ruben is correct, then *Ar-*

chaeopteryx was aerodynamically competent well before endothermy evolved. Thus selection for the evolution of endothermy in the avian lineage may have been effectively selection for flights of longer duration.

As always when the subject is *Archaeopteryx*, objections were expressed rapidly. J. R. Speakman—a zoologist at the University of Aberdeen who specializes in flight—questioned both the suitability of Pennycuick's model of flight demands and Ruben's calculation of the output of reptilian muscle. Pennycuick's model was inappropriate for this use, argued Speakman, because it estimates the mechanical power requirements of flight rather than the metabolic energy that had to be needed to supply that mechanical power. Muscles are rarely if ever 100 percent efficient in generating mechanical power; they may be arranged in a less-than-optimal position for any particular movement, for example. Thus apparently "excess" metabolic power must be generated in order to create the requisite mechanical power. After adjusting Pennycuick's model, Speakman reestimates the metabolic needs of a flying *Archaeopteryx* at more than two and a half times the maximum that a reptilian *Archaeopteryx* could generate with a flight-muscle ratio of 0.09. Speakman also questions the specific power output of 450 watts per kilogram that Ruben suggests is correct for reptiles. Although additional evidence has since accumulated that reptilian muscle indeed does generate more power than mammalian or avian muscle, the exact energy output of reptilian muscle must remain uncertain until additional physiological studies are completed.

Ruben's research did not directly address the requirements of takeoff in *Archaeopteryx*, except to say that a reptilian *Archaeopteryx* with half the flight muscle of a bird could take off as well as a bird. But he provided Marden with an estimate, however heuristically designed, of the minimal flight-muscle ratio of *Archaeopteryx*. This figure became useful in a further study of the requirements of takeoff, in which Marden found some subtle interactions among the factors involved in lift production. The maximum amount of lift depends directly on the mass of the flight muscles of the animal in question (with a marginal flight-muscle ratio of 0.16) and not upon differences in absolute body size. Nonetheless, takeoff and other activities requiring a short burst of power to produce lift *are* affected by absolute body size. This happens because animals of different sizes vary in the maximum lift they can produce per gram of muscle. As Marden explains, "Larger animals generate less lift per muscle power output. . . . In other words, larger animals must generate more power to achieve the same lift." The ratio of lift to power, as well as the ratio of flight muscle mass to total body mass, is crucial.

To take off, an animal must generate at least 9.8 newtons per kilogram (N/kg) of its body weight in order to overcome the effects of gravity. The amount of power that can be produced by a warm-blooded animal is also fairly standard: about 100 W/kg of muscle in aerobic metabolism or 225 W/kg in anaerobic metabolism. If the flight-muscle ratio is lower than 0.16, a bird cannot produce enough power to generate sufficient lift for takeoff using aerobic metabolism and must switch to anaerobic metabolism. But sustained flight cannot rely upon anaerobic metabolism or the flier will go into severe oxygen debt and exhaustion. The options that avoid this

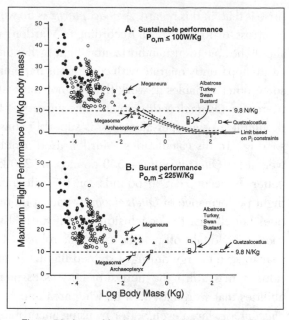

Figure 76. James Marden calculated the maximum flight performance of a 200-gram *Archaeopteryx* with a flight muscle ratio of 0.07. (A) Theoretically, a flier must produce at least 9.8 newtons of power per kilogram of body weight to sustain horizontal flight once aloft. In normal, aerobic metabolism, *Archaeopteryx* could produce only about half of this theoretical minimum, but its performance is comparable to that of several living birds that are awkward fliers. (B) In takeoff, anaerobic or burst-performance metabolism produces more power per unit of body mass. If it weighed 250 grams or less, *Archaeopteryx* could achieve takeoff with a flight-muscle ratio of only 7 or 8 percent, a value well below the minimum observed in most flying birds today.

fate are either to terminate the flight after a short distance or to use intermittent flapping strategies to minimize the energy expenditure.

With Ruben's estimate of a 0.07 flight-muscle ratio, Marden could then calculate the flight abilities of *Archaeopteryx*. His equations rely upon flight-muscle ratio, body weight (Marden used 200 grams), wingspan, and the maximum energy output for a warm-blooded animal under aerobic and anaerobic conditions as input. The output is the maximum flight performance (the lift per kilogram of body weight) of *Archaeopteryx,* compared to that of a range of living species and the pterosaur *Quetzalcoatlus* (Figures 76A and 76B). Using normal, aerobic metabolism, *Archaeopteryx* could produce only about half of the minimum lift required to sustain flight. The surprise is that, in this regard, *Archaeopteryx* is no worse off than a turkey, swan, albatross, or Kori bustard. According to Marden's model, none of these birds should be able to sustain horizontal flight by flapping for very long—although swans apparently migrate with continuous flapping, which suggests there are some other variables at work that are not understood at present. The giant pterosaur *Quetzalcoatlus* apparently suffered from this problem, too, although it, like a super-albatross, is superbly constructed to save energy by soaring. (In his calculations, Marden used a rather high estimate of body weight for *Quetzalcoatlus*—550 pounds, or 250 kilograms, versus the more generally accepted 280 pounds, or 127 kilograms—but recalculating the flight performance of *Quetzalcoatlus* with a more conservative body weight does not change the conclusions significantly. Even a 280-pound *Quetzalcoatlus* could take off readily, using anaerobic metabolism, but would have to soar to save energy during sustained flight.) In any case, Marden's calculations seem to affirm the more subjective assessment of *Archaeopteryx*'s aerial abilities that were based on its wing area and wing loading. In takeoff, *Archaeopteryx* fared even better by using anaerobic, short-burst power to produce slightly more than the minimum lift needed for takeoff, even with a very small flight-muscle ratio (Ruben's 0.07) and an avian physiology.

Marden used 200 grams as the body weight of *Archaeopteryx* in his calculations. Substituting the more generally accepted value of 250 grams changes the outcome. With a flight-muscle ratio of 0.07, a 250-gram *Archaeopteryx* would produce 9.0 N/kg under anaerobic conditions—just slightly less than the minimum needed for takeoff—and, in aerobic metabolism, too little to sustain flight. In this regard, though, *Archaeopteryx* is no worse off than a Kori bustard. However, these estimates are so close to the theoretical thresholds for flight that even a slight increase in the flight-muscle ratio would restore *Archaeopteryx* to flighted status. With a flight-muscle mass of 0.08—only slightly more than Ruben's arbitrary estimate of 0.07—a 250-gram *Archaeopteryx* could both take off and maintain sustained flight with ease. Remember that the flight-muscle ratio of 0.07 is an entirely arbitrary value, selected by Ruben to make a point. No one would suggest that this value is a precise estimate. Indeed, neither Ruben nor anyone else has found a convincingly secure way to estimate flight-muscle mass on *Archaeopteryx* precisely. Whether it was sheer coincidence or intuition, Ruben selected an arbitrary flight-muscle ratio very close to the minimum that would enable a 200–250-gram *Archaeopteryx* to fly.

Of course, calculating the flight performance of an extinct species is a quagmire of untested assumptions and theoretical pitfalls. But the real issue is not whether *Archaeopteryx* could have had a flight-muscle ratio of at least 0.08—which seems highly probable—but *why Archaeopteryx* could take off and fly with such a low flight-muscle ratio when living birds cannot. In making his calculations, Marden was surprised that neither an avian flight-muscle ratio (greater than 0.16) nor a reptilian physiology seemed to be necessary for flight in *Archaeopteryx*. After thinking the paradox over, Marden suggests two possible explanations. One lies in the details of his calculations. The first step is to calculate the lift-to-power ratio for the animal in question, from flight-muscle mass and wing length, which he derived from Ruben's and Yalden's papers. Their data made an *Archaeopteryx* with a very low flight-muscle mass relative to the length of its wings. In birds— and Marden's equations are derived from bird data—a low flight-muscle mass invariably indicates a low body mass. The false implication is that *Archaeopteryx* also had a large wingspan for its weight, which it does not in Yalden's reconstruction. But a large-winged bird with a low body weight will have a high lift-to-power ratio and, in the next step, the lift-to-power ratio is used to calculate the value for maximum flight performance. Of course, an inflated lift-to-power ratio will in turn inflate the maximum flight performance, making *Archaeopteryx* seem as if it could take off when perhaps it could not. A flight-muscle ratio of 0.07 places *Archaeopteryx* outside of the range of variability of the living animals from which the equations were originally derived. Possibly Marden's model makes good predictions only within the range of living birds.

And *Archaeopteryx* was not a modern bird. Neither Yalden's reconstruction nor Rietschel's fits the proportions of modern birds perfectly—and any *Archaeopteryx*, whatever its wing size and shape, surely had a low flight-muscle ratio. But Rietschel's reconstruction of *Archaeopteryx* is genuinely large-winged for its body weight, suggesting that the implications of a high lift-to-power ratio in Marden's calculations may not be false after all, if Rietschel's is the more accurate reconstruction. Marden's other explanation is an intriguing speculation, based on his keen awareness that things always change in evolution. Put simply, it is: "With an animal of entirely anomalous body shape, who knows if the same relationships apply?" He wonders if an animal with an entirely different set of bodily proportions—especially one with larger wings for its body size than most birds—might be able to escape from the avian rules of flight-muscle ratio and fly. Big wings may not always need a large motor.

With its large, broad wings and low wing loading, Rietschel's reconstruction of *Archaeopteryx* illuminates how flight could have evolved. His *Archaeopteryx* would have been capable of takeoff from the ground and sustained horizontal flight without any doubt, if its flight-muscle ratio was larger than 8 percent of its total body weight and if Marden's models are reasonably accurate. With the added advantage of modern, aerodynamically designed feathers, *Archaeopteryx* could apparently compensate for its small pectoral musculature, whether it was reptilian or avian in physiology. Refining its anatomy—enlarging its pectoral musculature, changing its shoulder structure to permit a wing flip, altering its proportions, and maybe even developing endothermy—could well have been later evolutionary advances that improved the flight of which *Archaeopteryx* was already capable.

There is also a satisfying sensibleness about concluding that *Archaeopteryx* could and did fly in a vegetated habitat, although it could take off from the ground. In evolving flight, it is only reasonable to believe that *Archaeopteryx* adapted to exploit a vacant, aerial niche—one not already occupied by the pterosaurs whose fossils are also preserved at Solnhofen. Pterosaurs are built like soaring animals, specialized for open habitats and probably (in many cases) open water, some of them seemingly spearing fish near the surface and others—as has been suggested for *Quetzalcoatlus*—landing along lake shores and digging for arthropods and other invertebrates, using their long beaks or claws. With their long wings and high aspect ratios, pterosaurs were poorly adapted to flying in vegetated or closed habitats that would demand high maneuverability. What pterosaurs couldn't do and *Archaeopteryx* probably could was fly among the shrubs and ground cover, chasing insects and maybe even catching them on the wing, as flycatchers do today. Evolving *Archaeopteryx* was just a matter of taking the predatory behavior of a ground-based animal and finding a way to carry that behavior up into the air. But now the focus must be on the steps between theropods, the most probable ancestor of birds, and *Archaeopteryx*. "Proavis" is still a mystery.

✦   ✦   ✦

Coupling Rietschel's new reconstruction with our newly improved understanding of the prerequisites of flight resolved many of my questions about

Archaeopteryx and the origins of bird flight. I am now convinced that *Archaeopteryx* was such a large-winged creature that it could take off from the ground, with either a reptilian or an avian physiology—and I shall follow the developing evidence about its physiology with keen interest. The key to understanding *Archaeopteryx* is recognizing that it was not a bird as birds are today, but an evolutionary fledgling. We should not have expected otherwise. I do not fool myself, however, that the debates are over or that the arboreal proponents will suddenly abandon their position. Still, Rietschel's reconstruction seems right to me, and it is most congruent with (although not dependent upon) dinosaurian ancestry for *Archaeopteryx.* During the researching and writing of this book, I have been transformed slowly from an outsider, struggling to understand the issues, into a researcher exploring a few avenues that had been overlooked by others, and finally into an adherent of the "ground up" hypothesis. Still, I have endeavored to present a balanced and fair review of the current state of the controversy—and have tried to point out strengths and weaknesses wherever they occur.

Whatever else is shown by this not-always-academic conflict, it is clear that methodology is key in attempting to resolve these complex arguments and vexing uncertainties. Convictions, beliefs, and personal prejudices can and do inspire researchers to think up new projects and new techniques; this is all for the good. No researchers ever slogged away, meticulously taking hundreds of tedious measurements or making dozens of detailed anatomical comparisons, because the tasks in and of themselves are fun; they have done this because the data they would gain promise to yield an answer to a question that matters to them personally.

But opinion and personal satisfaction cannot be used to evaluate the relative merits of competing theories or the accuracy of different models of reality, unless we are speaking of religion and not science. What is theoretically logical or more appealing counts for little in the final analysis, for, as Kevin Padian has said, "Nature sometimes has a way of not following the easiest path." The fossil record of avian evolution is a tangled wing. Disentangle it properly and you can suddenly fly and soar to new heights, gaining new perspectives and understandings. Leave it tangled and your attempts at flight are bound to end in disastrous crashes.

The essence of science is that theories and models must be *tested,* rigorously—tested against the mute testimony of the fossils and their features and tested against the implacable physical laws of our universe. *Archaeopteryx* could not fly by revving up a gasoline engine, because it didn't have one, nor could it take off by banishing gravity, which it could not do. Whatever its biology and life were like, *Archaeopteryx* had to evolve within a world that operates within rules much like those of today, even if the inhabitants of that world were rather different from those around us now.

And the testing process is ongoing, the assessment ever-changing. As I write, additional evidence of the detailed anatomical resemblances between dinosaurs and birds is mounting. Furculas have been found in more dinosaurs; additional dinosaurs are being scanned with computed tomography to check the sizes and shapes of their nasal cavities and the presence or absence of ridges for the attachment of nasal turbinates; experimental biologists are documenting the details of flight ever more closely. But the most stunning finding of all was reported informally at the Society of Vertebrate Paleontology meetings in October 1996. There, Philip Currie of the Royal Tyrrell Museum of Palaeontology in Drumheller, Alberta, and Chen Pei-Ji of the Nanjing Institute of Geology and Paleontology in China shared some pictures of a startling new find with their professional colleagues. From the very same freshwater lake deposits in Liaoning Province in northeastern China that yielded the oldest beaked bird, *Confuciusornis*—believed to be close in time to *Archaeopteryx*—has come evidence of a small theropod dinosaur *with feathers* (Figure 77).

Feathered dinosaurs hit the front pages of newspapers around the world. Indeed, as Currie says, "In normal, sane times, nobody would release [a

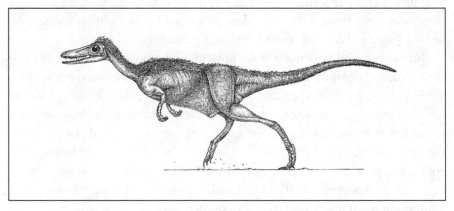

Figure 77. Late in 1996, the discovery of a remarkable small theropod dinosaur, *Sinosauropteryx prima*, was announced by Chinese scientists. The specimen is very similar to *Compsognathus* from Solnhofen, but it has some sort of hairy or downy fibers covering the back from head to tail, as shown here. Additional study soon revealed that these fibers also extend down the sides of the body and along the arms and legs. If the fibers are a type of down or other insulation, this nonflying species was probably endothermic. The exact nature of the fibers is yet to be determined.

find like this] to the press before they had a scientific paper ready to publish." But the Chinese press had already talked to Dr. Ji Qiang, director of the Chinese Geology Museum in Beijing, who named the specimen as the type of a new species, *Sinosauropteryx prima,* or the first Chinese dinosaur-wing, on the basis of very preliminary studies. By the time of the Society of Vertebrate Paleontology meetings, the story was filtering through to the North American media, so Currie and Chen decided it was appropriate to let fellow paleontologists see the photograph and sketch of the fossil.

What is preserved is the skeleton of a small theropod, very similar to *Compsognathus,* about one meter (39 inches) in total length. Along the top of the skull and down the spine to the tip of the tail, and extending down the body, arms, and legs, is what Currie calls "a sort of halo" that looks like down or feathers. In fact, this tissue may have covered the entire body. Currie hasn't yet examined the tissue under a good microscope. He says,

> I was forewarned that they had a feathered dinosaur and the Chinese fundamentally invited me over to take a look at it. I had been told there were feather impressions and had been shown a photograph, but I was expecting to see dendrites [a branching mineral deposit common on fossils] or something. Within 10 seconds, that was out of my mind and I knew what I was looking at. They are soft, pliable tissues of some kind. You can see they had been clumping together before the carcass was buried—this stuff clumps up like wet feathers or wet hair, so it is hard to make out just what's going on. But in the tail region, you can see them as independent entities. Each looks sort of like a scale that has been subdivided into much finer parts or like a down feather where the rachis is offset to one side completely. I really think they are individual feathers or scales; it's the most parsimonious explanation.
>
> I think this specimen will go a long way toward convincing the skeptics of the existence of hot-blooded dinosaurs. You can argue whether these are feathers or feather-like scales [until the microscopic studies have been completed]. But this tissue is there for a purpose and it's hard to imagine what it's for if not for insulation. My intuition is that these are going to be feathers, but there's a difference between what your intuition is and what one should say as a scientist.

If this tissue proves to be feathers or an evolutionary precursor to feathers, then the "ground up," dinosaurian ancestry theory will receive a major boost in credibility. *Sinosauropteryx* is clearly bipedal and clearly has arms, not wings. Thus, if it possesses unmistakable feathers, this fossil would establish with certainty that feathers could and did evolve in nonflying theropod dinosaurs. This fact would demolish Feduccia's argument that feathers are intrinsically and fundamentally aerodynamic structures.

It is particularly ironic that *Sinosauropteryx* so closely resembles the one species of theropod, *Compsognathus,* with which *Archaeopteryx* has been most commonly confused. John Ostrom once searched the Solnhofen *Compsognathus* specimens with care, looking in vain for feather traces. He wrote:

> The reader can be sure that I made an exhaustive examination, . . . but to no avail. If feathers had been present in *Compsognathus,* it is inconceivable to me that no evidence of them would be preserved. . . . Thus, I conclude that *Compsognathus* almost certainly was not feathered.

Now, from a site halfway around the world, comes contradictory evidence of feather traces on a theropod, even if that particular theropod is too recent to be the direct ancestor of *Archaeopteryx.*

As with the initial discovery of *Archaeopteryx,* the fossil record has—remarkably—yielded up a telling piece of evidence just as the objections to the cursorial theory have gotten louder. In his recent book, *The Origin and Evolution of Birds,* Alan Feduccia argued strongly against the idea of hot-blooded dinosaurs, particularly singling out the question of feathers. Under a reconstruction of a feathered theropod, Feduccia wrote, "There is no evidence that any dinosaur possessed feathers; feathers are absolutely unique to birds," which was, of course, completely true at the time. Asked about the newspaper reports of the feathered dinosaur, Feduccia was dismissive. He had not yet seen the photographs of the specimen nor read any scientific presentation of the evidence (for none was available). He said,

> The problem with the thing is that feathers are basically aerodynamic. Presumably these people want this dinosaur to be insulated, endothermic. Well, a lot of these theropods have ornamentation down the back, but not feathers. . . . If it's feath-

ers, it's down. Down does not give rise to feathers, but just the opposite; down is secondary in modern birds. It just makes no sense to me.

He is certain that, when more closely examined, the structures will prove not to be feathers. But, however unanticipated, however illogical, and however incredible it is to some, the fact is that this small theropod has feathers or something very much like them. And they are obviously thermoregulatory in function, for *Sinosauropteryx* has no wings. And a new theropod dinosaur from Argentina, *Unenlagia*, has arms that fold up against its body, like a bird's wings. To Currie and others, the evidence for a hot-blooded dinosaur is now "overwhelming." Even to those on the fence, the case for the "ground up" hypothesis of avian flight has suddenly grown much, much stronger. Unless and until a good, nondinosaurian candidate for avian ancestry is discovered, the "trees down" hypothesis will remain a minority view with the weight of evidence against it.

The enduring benefit of the debate is that it forces proponents of both theories to seek out new evidence, to examine their own (and others') assumptions closely, and—where appropriate—to improvise new types of analyses and studies. I have come to realize that the story of the evolution of *Archaeopteryx* and of the evolution of the interpretation of *Archaeopteryx* are parallel tales. *Archaeopteryx* itself, whatever its ancestry, did not evolve with intent to become a bird or a flying animal. It responded to opportunities, making use of vacant ecological niches and anatomical potentials as it could. Evolution in this case, as in so many others, was really a matter of winging it—of improvising, of doing the best that circumstances allowed. Similarly, the paleontologists, biologists, physiologists, aerodynamic engineers, and ornithologists struggling to understand this wonderful fossilized masterpiece have been forced to invent their science as they have gone along. Before the discovery of the first skeleton of *Archaeopteryx*, no scientist had considered what a truly transitional form would look like or how its anatomy was to be analyzed. The fossil itself, by its magnificent preservation and by its tantalizing combination of the familiar and the strange, has spurred researchers to develop new techniques, new modes of comparison, and a new appreciation of the laws and principles of physics, physiology, embryology, and functional anatomy.

The intractable problems and nearly endless questions posed by *Archaeopteryx* have drawn together an eclectic array of workers from many different backgrounds. Each specialist has been pressed for insights, squeezed for data, and begged for laws, rules, or approaches to analysis that can be borrowed to resolve the tricky issues at hand. That this remarkable group has struggled together is surprising; that they have often disagreed is not. Their perspectives are wildly divergent and their underlying assumptions are often at odds with each other; even the levels of precision with which they measure are different.

It has been, at best, a process of mutual education: paleontologists reaching out to experts in fields remote from their own, seeking wisdom, followed by pronouncements from the experts that must later be tempered as they, in turn, become more educated about fossils and evolutionary processes. The process is iterative. Someone gets an idea that promises to resolve an important issue: Were dinosaurs warm-blooded? Were birds descended from dinosaurs? Could *Archaeopteryx* fly? Did theropods have feathers? But what at first seems a clinching piece of evidence—an irrefutable test of one or the other hypothesis—is later neither so certain nor so secure.

Part of the difficulty lies in the fact that the past was *not* like the present. Dinosaurs aren't birds—at least, not birds as they are now—and the first feathers weren't feathers as they are now, either. Even *Archaeopteryx*, the hallowed First Bird, isn't a bird as birds are now; it is only closer to them than it is to anything else we can lay our hands on. Explorers of the past, like explorers of a new continent, seek the familiar to ground themselves; it is an essentially human and nearly irresistible impulse. But the essence of evolution is that all things change: species, habitats, structures, and functions. What is admirable is that, on the whole, those who have disagreed have remained respectful of their opponents and willing to listen to divergent views. Almost every scientist involved in these debates has asserted his or her willingness to listen to new evidence, to abandon a dearly held hypothesis if it is proven false. But abandonment rarely occurs, especially when men and women have invested their lives and research in exploring a particular point of view. Not everyone has the courage to face evidence that challenges a well-constructed and long-held theory with aplomb, much less to dive in enthusiastically and pursue a new direction. Understanding the past, like evolving flight, is not an easy goal to accomplish. Scientists are human, too, and humans have long been fond of the familiar and wary of the strange. The struggle is to keep the focus not on *being right* personally but on *finding out* the right answer.

Finally—always—there are the specimens of *Archaeopteryx* themselves, all that is left of a wonderful, enigmatic creature that took to the skies. That the specimens are beautiful is a gift to those who view them. Their elegance, their unique place in the history of the acceptance of evolutionary theory, and their charm as portraits of the First Bird make *Archaeopteryx* very special. To some extent, these qualities hinder clear thinking and objective treatment of the fossils. But they also inspire passionate fascination—a deep concern with the truth—from which springs the best kind of science. Like more fossils of *Archaeopteryx*, the truth is waiting to be discovered.

Notes

Chapter 1. Taking Wing

22 It was part of a collection belonging to Carl Häberlein: W. T. Stearn, *The Natural History Museum at South Kensington; A History of the British Museum (Natural History) 1753–1980*, 1981; J. Ostrom, "Introduction to *Archaeopteryx*," in M. K. Hecht, J. Ostrom, G. Viohl, and P. Wellnhofer, eds., *The Beginnings of Birds*, 1985; G. Viohl, "Carl F. and Ernst O. Häberlein, the sellers of the London and Berlin specimens of *Archaeopteryx*," in M. K. Hecht, et al., eds., *The Beginnings of Birds*, 1985.

25 "absence or rarity . . .": C. Darwin, *The Origin of Species by Means of Natural Selection or the Preservation of Favored Races in the Struggle for Life*, 1859, reprinted 1958, p. 158.

25 "But," Darwin writes: Ibid., p. 159.

25 "The explanation lies, as I believe . . .": Ibid., pp. 291ff.

26 "In conclusion, I must add . . .": J. A. Wagner, "Über ein neues, angeblich mit Vögelfedern versehenes Reptil aus dem Solnhofener lithographischen Schiefer," *Sitzungsber. Bayer. Akad. Wiss.*, 1861, translated by W. S. Dallas, "On a new fossil reptile supposed to be furnished with feathers," *Ann. Mag. Nat. Hist.*, 1862.

26 Together with the rest of Häberlein's collection: W. T. Stearn, *Natural History Museum*, 1981.

26 In 1862, Thomas Huxley: A. Desmond, *Huxley: The Devil's Disciple*, 1994, p. 322.

281

26 In Victorian England, the number of servants: I. Beeton, *The Book of House-hold Management,* 1861, p. 7.

27 Darwin came from a far wealthier family: A. Desmond and J. Moore, *Darwin,* 1991.

27 "ambitious, very envious . . .": Quoted in L. Barber, *The Heyday of Natural History: 1820–1970,* 1980, pp. 170ff.

28 Even the mild-mannered Charles Darwin: F. Darwin, ed., *Charles Darwin: His Life Told in an Autobiographical Chapter and in a Selected Series of His Published Letters,* 1892, reprinted 1958, p. 61.

28 "unequivocally a bird," "a closer adhesion . . .": R. Owen, "On the *Archaeopteryx* of von Meyer, with a description of the fossil remains of a long-tailed species, from the lithographic stone of Solnhofen," *Phil. Trans. Roy. Soc. Lond.,* 1863.

29 "The unique specimen . . . ," "It is obviously impossible . . .": T. H. Huxley, "Remarks upon *Archaeopteryx lithographica,*" *Proc. Roy. Soc. Lond.,* 1868, pp. 243–44.

30 "The soft tortoises . . .": Ibid., p. 248.

31 "1. Are any fossil birds . . .": T. H. Huxley, "On the animals which are most nearly intermediate between the dinosaurian reptiles and birds," *Sci. Mem.,* 1868, vol. 3, p. 307. Quoted in M. di Gregorio, *T. H. Huxley's Place in Natural Science,* 1984, pp. 83–84.

31 "an air of pulling rabbits . . .": M. di Gregorio, *Huxley's Place,* 1984, p. 81.

35 "I find [he writes] . . .": T. H. Huxley, "Further evidence of the affinity between the dinosaurian reptiles and birds," *Quart. J. Geol. Soc. Lond.,* 1870, pp. 30–31.

37 Working through a friend in Dresden: This story is told in J. Ostrom, "The Yale *Archaeopteryx*: The one that flew the coop," in M. K. Hecht, et al., eds., *The Beginnings of Birds,* 1985.

39 "the most beautiful museum . . .": J. de Vos, pers. comm. to author, 1994.

40 "museum of a museum": J. de Vos, pers. comm. to author, 1994.

40 Ostrom had come to study: The account that follows, including quotations, is taken from J. Ostrom, interview with author, 1/20/95.

Chapter 2. What's the Flap?

47 "He is The Man": D. B. Weishampel, interview with author, 1995.

48 "This issue of bird flight . . .": J. Ostrom, interview with author, 1/20/95.

48 "To attain to the true science . . .": Leonardo da Vinci, Ms. E., fol. 54r and *Codex Atlanticus* 45r, quoted in L. B. Hart, *The World of Leonardo da Vinci: Man of Science, Engineer and Dreamer of Flight,* 1961.

49 About one hundred years later, John Wilkins: *Encyclopaedia Britannica,* "Flight, history of," *Macropaedia,* vol. 7, 1974.

56 Ancient Japanese feather armor: J. E. Gordon, *Structures; or Why Things Don't Fall Down,* 1978, p. 130.

57 Even as some birds twist their wings: R. Å. Norberg, "Function of vane asym-

metry and shaft curvature in bird flight feathers; inferences on flight ability of *Archaeopteryx*," in M. K. Hecht, et al., eds., *The Beginnings of Birds*, 1985.

60 the radius acts as a mechanical connecting rod: The best technical discussion of this mechanism occurs in R. Vasquez, "Functional osteology of the avian wrist and the evolution of flapping flight," *J. Morphol.*, 1992.

62 The muscle that is primarily responsible: See discussion in Vasquez, op. cit.; J. Ostrom, "Some hypothetical anatomical stages in the evolution of avian flight," S. L. Olson, ed., *Smithson. Contrib. to Paleobiol.*, 1976; E. Coues, "On the mechanism of flexion and extension in birds' wings," *Proc. Amer. Assn. Adv. Sci.*, 1871; H. L. Fisher, "Bony mechanisms of automatic flexion and extension in the pigeon's wing," *Science*, 1957.

64 "to make a surface . . .": *Encyclopaedia Britannica*, "Flight, history of," *Macropaedia*, vol. 7, 1974, p. 384.

64 Otto Lilienthal: This discussion of Lilienthal, the Wright brothers, and early aviation is taken largely from F. Howard, *Wilbur and Orville: A Biography of the Wright Brothers*, 1987, and R. Wohl, *A Passion for Wings: Aviation and the Western Imagination, 1908–1918*, 1994.

65 "Sacrifices must be made": F. Howard, *Wilbur and Orville*, 1987, p. 16.

66 Wilbur always claimed: Ibid., p. 33.

Chapter 3. Flight Plan

68 "One evening, as we were going . . .": K. Blixen, *Out of Africa*, 1937, reprinted 1973, pp. 327–28.

69 One of the first and most famous: M.-H. Sy, "Funktionell-anatomische Untersuchungen am Vögelflugel," *J. Orn., Lpz.*, 1936.

70 I turned to an old friend: The report of this work is given in F. A. Jenkins, Jr., K. P. Dial, and G. E. Goslow, Jr., "A cineradiographic analysis of bird flight: The wishbone in starlings is a spring," *Science*, 1988; G. E. Goslow, Jr., K. P. Dial, and F. A. Jenkins, Jr., "Bird flight: Insights and complications," *BioScience*, 1990.

75 during flight, starlings show: G. E. Goslow, Jr., et al., "Bird flight," 1990, p. 113.

77 "This is the reason . . .": C. Pennycuick, "Mechanical constraints on the evolution of flight," in K. Padian, ed., *The Origin of Birds and the Evolution of Flight*, 1986, p. 95.

77 "Bats use a different method . . .": Ibid., p. 95.

78 The next step in the intricate tango: D. F. Boggs and K. P. Dial, "Neuromuscular organization and regional EMG activity of the pectoralis in the pigeon," *J. Morphol.*, 1993; S. M. Gatesy and K. P. Dial, "Tail muscle activity patterns in walking and flying pigeons (*Columbia livia*)," *J. exp. Biol.*, 1993; K. P. Dial, "Activity patterns of the wing muscles of the pigeon (*Columbia livia*) during different modes of flight," *J. exp. Zool.*, 1992.

80 *birds don't use their wing muscles*: K. P. Dial, "Avian forelimb muscles and nonsteady flight: can birds fly without using the muscles in their wings?," *Auk*, 1993.

80 "a consequence of the fact . . .": Ibid., p. 881.

82 Another energy-saving mechanism: C. Pennycuick, *Bird flight performance; a practical calculation manual,* 1989; C. Pennycuick, "Soaring behavior and performance of some East African birds, observed from a motor-glider," *Ibis,* 1972; Å. Hedenström, "Migration by soaring or flapping flight in birds: the relative importance of energy cost and speed," *Phil. Trans. Roy. Soc. Lond.,* 1993.

83 Gatesy and Dial have proposed: S. M. Gatesy and K. P. Dial, "Locomotor modules and the evolution of avian flight," *Evol.,* 1996.

86 their tails have yet another way of functioning: S. M. Gatesy and K. P. Dial, "Tail muscle activity," 1993.

Chapter 4. Nesting Sites

92 Starting with *Archaeopteryx:* See reviews by L. Chiappe, "The first 85 million years of avian evolution," *Nature,* 1995, and A. Feduccia, "Explosive evolution in Tertiary birds and mammals," *Science,* 1995.

93 recent discoveries have complicated: L. D. Martin, "The origin and early radiation of birds," in A. H. Brush and G. A. Clark, Jr., eds., *Perspectives in Ornithology,* 1983; L. D. Martin, "The beginning of the modern avian radiation," in C. Mourer-Chauviré, ed., *L'Évolution des Oiseaux d'après le témoignage des fossiles,* 1987; A. Feduccia, *The Origin and Evolution of Birds,* 1996; L. Chiappe, "The first 85 million years," 1995.

94 So does an unnamed: J. L. Sanz, L. M. Chiappe, B. P. Pérez-Moreno, J. J. Moratalla, F. Hernández-Carrasquilla, A. D. Buscalioni, F. Ortega, F. J. Poyato-Ariza, D. Russkin-Gutman, and X. Martínez-Delclòs, "A nesting bird from the Lower Cretaceous of Spain," *Science,* 1997.

94 "startling": P. Altangerel, M. A. Norell, L. M. Chiappe, and J. M. Clark, "Flightless bird from the Cretaceous of Mongolia," *Nature,* 1993, p. 623.

94 "similar to that of digging animals," "The short forelimb . . .": Ibid., p. 625.

95 a number of well-known figures: L. D. Martin, interview with author, 12/1/94; J. Ostrom, interview with author, 1/20/95; S. Olson, cited in A. Feduccia, *Origin and Evolution,* 1996, p. 86; A. Feduccia, *Origin and Evolution,* 1996, p. 86.

95 "The placement of *Mononykus . . .*": L. Chiappe, "The first 85 million years," 1995, p. 352.

96 "This meant that on land . . .": L. Martin, "Origin and early radiation," 1983, pp. 314–15.

97 "The thing that strikes me . . .": This remark recounted by A. Feduccia during an interview with author, 11/20/96.

98 "letters to all the natural history museums . . .": J. R. Horner and J. Gorman, *Digging Dinosaurs,* 1988, p. 22. Other details of Horner's life and work are also taken from this source.

98 "look interesting": Ibid., p. 21.

99 "If this was true . . .": Ibid., pp. 58–59.

99 The evidence from this nest: J. R. Horner and R. Makela, "Nest of juveniles provides evidence of family structure among dinosaurs," *Nature,* 1979.

99 Exactly who first thought: Who deserves priority for the warm-blooded dinosaur idea is a contested point and the cause of bad feelings among a number of dinosaurologists. Some of the important and early papers exploring various aspects of the idea are: R. T. Bakker, "The superiority of dinosaurs," *Discovery,* 1968; R. T. Bakker, "Dinosaur physiology and the origin of mammals," *Evol.,* 1971; R. T. Bakker, "Anatomical and ecological evidence of endothermy in dinosaurs," *Nature,* 1972; J. Ostrom, "Terrestrial vertebrates as indicator of Mesozoic climates," *N. Amer. Paleo. Conv. Proc.,* 1970; A. J. de Ricqlès, "Evolution of endothermy: Histological evidence," *Evolutionary Theory,* 1974.

100 "This was the first time . . .": J. R. Horner and J. Gorman, *Digging,* 1988, p. 63.

101 the first theropod embryos ever found: M. A. Norell, J. M. Clark, D. Demberelyin, B. Rhinchen, L. M. Chiappe, A. R. Davidson, M. McKenna, P. Altangerel, M. J. Novacek, "A theropod dinosaur embryo and the affinities of the Flaming Cliffs dinosaur eggs," *Science,* 1994; press release, American Museum of Natural History, 12/20/95; see also M. A. Norell, J. M. Clark, R. Weintraub, L. M. Chiappe, and D. Demberelyin, "A nesting dinosaur," *Nature,* 1995.

102 "The very first thing you do . . .": L. D. Martin, interview with author, 12/1/94, for this quote and other opinions expressed in this paragraph.

102 "I don't give a damn . . .": J. R. Horner, interview with author.

103 A simple list: For discussion, see J. Ostrom, "*Archaeopteryx* and the origin of flight," *Quart. Rev. Biol.,* 1974; J. Ostrom, "The question of the origin of birds," in H.-P. Schultze and L. Trueb, eds., *Origins of the Higher Groups of Tetrapods,* 1991.

105 The most inflammatory issue of all: This discussion owes much to a review by M. Benton, "Origin and interrelationships of dinosaurs," in D. B. Weishampel, P. Dodson, and H. Osmolska, eds., *The Dinosauria,* 1990. However, there is no general consensus about the details of dinosaur cladistics.

106 "In current usage . . .": J. Gauthier and K. Padian, "Phylogenetic, functional, and aerodynamic analyses of the origin of birds and their flight," in M. K. Hecht, et al., eds., *The Beginnings of Birds,* 1985, p. 189.

109 "Birds are as much . . .": J. Gauthier, quoted in C. Zimmer, "Ruffled feathers," *Discover,* 1992, p. 48.

109 The history of theories: For an intelligent review, see L. Witmer, "Perspectives on avian origins," in H.-P. Schultze and L. Trueb, eds., *Origins of the Higher Groups of Tetrapods,* 1991.

109 The initial discovery of *Archaeopteryx:* E. D. Cope, "An account of the extinct reptiles which approached the birds," *Proc. Acad. Natl. Sci. Philadelphia,* 1867; T. H. Huxley, "On the animals which are most nearly intermediate between birds and reptiles," *Ann. Mag. Nat. Hist.,* 1868; T. H. Huxley, "Further evidence of the affinity between the dinosaurian reptiles and birds," *Quart. Geol. Soc. Lond.,* 1870.

110 Robert Broom, an irrepressible: R. Broom, "On the South African pseudosuchian *Euparkeria* and allied genera," *Proc. Zool. Soc. Lond.,* 1913; R. Broom,

"On the early development of the appendicular skeleton of the ostrich, with re-marks on the origin of birds," *Trans. S. Afr. Phil. Soc.*, 1906.

110 a very influential book: G. Heilmann, *Origin of Birds*, 1926.

110 "This quotation illustrates . . .": L. Witmer, "Perspectives," 1991, pp. 439–40.

111 "nothing in their structure . . .": G. Heilmann, *Origin of Birds*, 1926, p. 191.

111 *"Birds originated from bipedal . . ."*: F. Nopsca, "Ideas on the origin of flight," *Proc. Zool. Soc. Lond.*, 1907, p. 234.

112 a possible fossil bird: The scientific account of *"Protoavis"* is S. Chatterjee, "Cranial anatomy and relationships of a new Triassic bird from Texas," *Phil. Trans. Roy. Soc. Lond.*, 1991; for a popular account of the controversy, see C. Zimmer, "Ruffled feathers," 1992.

112 "I never went to the press . . .": S. Chatterjee, quoted in C. Zimmer, "Ruffled feathers," 1992, pp. 50–51.

112 "smushed and mashed and broken": J. Gauthier, quoted in Zimmer, op. cit., p. 50.

112 "real roadkill": T. Rowe, quoted in Zimmer, op. cit., p. 53.

112 Even Larry Martin: L. D. Martin, interview with author, 12/1/94.

113 "I have to admit . . .": L. Witmer, 3/26/96 E-mail to author.

113 The best course: L. Chiappe, "The first 85 million years," 1995.

113 Then, in 1972, Alick Walker: A. D. Walker, "New light on the origin of birds and crocodiles," *Nature*, 1972; A. D. Walker, "Evolution of the pelvis in birds and dinosaurs," in S. Andrews, R. Miles, and A. D. Walker, eds., *Problems in Vertebrate Evolution*, 1974.

113 "My viewpoint on the origin . . .": L. D. Martin, interview with author, 12/1/94.

114 In 1969, in Montana: J. Ostrom, "Terrible claw," *Discovery*, 1969.

114 Ostrom began to revive: J. Ostrom, "The ancestry of birds," *Nature*, 1973; J. Ostrom, "Origin of flight," 1974; J. Ostrom, *"Archaeopteryx* and the origin of birds," *Biol. J. Linnean Soc.*, 1976; see also J. Ostrom, "On the origin of birds and of avian flight," in R. S. Spencer, ed., *Major Features of Vertebrate Evolution*, 1994.

114 "Were it not for the preserved impressions . . .": J. Ostrom, "Origin of flight," 1974, p. 42.

115 Martin in particular has challenged: L. D. Martin, "Modern avian radiation," 1987; L. D. Martin, "The relationship of *Archaeopteryx* to other birds," in M. K. Hecht, et al., eds., *The Beginnings of Birds*, 1985; L. D. Martin, "The origin of birds and early flight," *Curr. Ornithol.*, 1983.

115 One of the weaknesses: L. Witmer, "Perspectives," 1991, p. 460.

116 "Oh, yes!": J. Ostrom, interview with author, 1/20/95; L. D. Martin, interview with author, 12/1/94.

Chapter 5. A Bird in the Hand

118 "attributes of two organisms . . .": Quoted in E. Mayr, *The Growth of Biological Thought: Diversity, Evolution and Inheritance*, 1982, p. 465, no citation given.

120 Nonhomologous parts cannot be evidence: J. R. Hinchcliffe, "'One, two, three' or 'Two, three, four': An embryologist's view of the homologies of the digits and carpus of modern birds," in M. K. Hecht, et al., eds., *The Beginnings of Birds,* 1985; S. Tarsitano, "*Archaeopteryx*—Quo Vadis?," in H.-P. Schultze and L. Trueb, eds., *Higher Groups,* 1991; L. D. Martin, "Mesozoic birds and the origin of birds," in H.-P. Schultze and L. Trueb, op. cit.

121 the automatic folding and unfolding mechanism: This mechanism is described in R. Vasquez, "Functional osteology," 1992.

121 "incapable of executing . . .": Ibid., p. 266.

122 "It was back in the 1970s . . .": This and following quotes from A. Feduccia, interview with author, 10/20/96.

123 "If these animals . . .": J. Ostrom, "Bird flight: How did it begin?," *Amer. Sci.,* 1979, pp. 54–55.

124 "The suggestion that the feathers . . .": L. D. Martin, "Origin of birds," 1983, p. 122.

124 according to careful reconstructions: S. Burkhard, "Remarks on reconstruction of *Archaeopteryx* wing," in M. K. Hecht, et al., eds., *The Beginnings of Birds,* 1985; S. Rietschel, "Feathers and wings of *Archaeopteryx,* and the question of her flight ability," in Hecht, op. cit. (see especially figure 4).

125 a trio from Northern Arizona University: G. Caple, R. Balda, and W. Willis, "The physics of leaping animals and the evolution of preflight," *Amer. Nat.,* 1983; also, R. Balda, G. Caple, and W. Willis, "Comparison of the gliding to flapping sequence with the flapping to gliding sequence," in M. K. Hecht, et al., eds., *The Beginnings of Birds,* 1985.

125 using small or incipient wings: K. Padian, "Running, leaping, lifting off," *The Sciences,* 1982.

126 Another contested point: M. E. Howgate, "*Archaeopteryx*—no new finds after all," *Nature,* 1983; M. E. Howgate, "*Archaeopteryx*'s morphology," *Nature,* 1984; R. A. Thulborn and T. L. Hamley, "The reptilian relationships of *Archaeopteryx,*" *Aust. J. Zool.,* 1982; J. R. Hinchcliffe and M. K. Hecht, "Homology of the bird wing skeleton: embryological versus paleontological evidence," *Evol. Biol.,* 1984; J. R. Hinchcliffe, "'One, two, three,'" 1985; M. K. Hecht and S. Tarsitano, "Paleontological myopia," *Nature,* 1984; M. K. Hecht and S. Tarsitano, "The paleobiology and phylogenetic position of *Archaeopteryx,*" *Geobios Mem. Sp.,* 1982; S. Tarsitano and M. K. Hecht, "A reconsideration of the reptilian relationships of *Archaeopteryx,*" *Zool. J. Linn. Soc.,* 1980.

127 detailed embryological evidence: Reviewed in J. R. Hinchcliffe, "'One, two, three,'" 1985.

127 After warm and lengthy debate: S. Tarsitano, "Quo Vadis?," 1991, pp. 566–67.

128 A similar problem pertains: L. D. Martin, J. D. Stewart, and K. Whetstone, "The origin of birds: structure of the tarsus and teeth," *Auk,* 1980; L. D. Martin and J. D. Stewart, "Homologies in the avian tarsus," *Nature,* 1985; L. D. Martin, "Mesozoic birds," 1991.

128 This discussion spurred Chris McGowan: C. McGowan, "Evolutionary relation-

ships of ratites and carinates: Evidence from ontogeny of the tarsus," *Nature,* 1984; C. McGowan, "Homologies in the avian tarsus, McGowan replies," *Nature,* 1985.

130 John Ostrom interprets: J. Ostrom, "Question," 1991, especially pp. 476ff.; L. D. Martin, interview with author; S. Tarsitano, "The morphological and aerodynamic constraints on the origin of avian flight," in M. K. Hecht, et al., eds., *The Beginnings of Birds,* 1985; A. Feduccia, "Evidence from claw geometry indicating arboreal habits of *Archaeopteryx,*" *Science,* 1993, p. 792; see also S. Tarsitano, "Constraints," 1985, and L. D. Martin, "Mesozoic birds," 1991.

130 Ostrom believes that at least some of them: J. Ostrom, "Question," 1991.

131 But Martin and Tarsitano put a different construction: L. D. Martin, "Mesozoic birds," 1991, p. 528; S. Tarsitano, "Quo Vadis?," 1991, p. 566.

131 dinosaur footprints, as did Baron Nopsca: F. Nopsca, "On the origin of flight in birds," *Proc. Zool. Soc. Lond.,* 1923.

132 "[If] a quadrupedal flying stage . . .": J. Ostrom, "Origin of flight," 1974, p. 30.

133 "Advocates of the arboreal theory . . .": J. Ostrom, "Origin of flight," 1974, p. 36.

133 First, the hallux in both *Archaeopteryx:* W. Bock and W. D. Miller, "The scansorial foot of the woodpeckers with comments on the evolution of perching and climbing feet in birds," *Am. Mus. Nat. Hist. Novitates,* 1959, p. 31; also, J. Ostrom, "Origin of flight," 1974, pp. 35ff.

135 In *Archaeopteryx,* the flexor tubercle: J. Ostrom, "Origin of flight," 1974, p. 37.

135 Feduccia quantified the geometry of the claws: A. Feduccia, "Claw geometry," 1993.

136 ". . . raptors (predatory birds), long-legged marsh birds . . .": Ibid., p. 790.

137 "strictly a perching adaptation . . .": A. Feduccia, quoted in V. Morell, "*Archaeopteryx*: Early bird catches a can of worms," *Science,* 1993, p. 764.

Chapter 6. Birds of a Feather

141 "life having evolved by blind chance . . .": C. Wickramasinghe, quoted in J. Mitchell, "When feathers fly: A case of fossil forgery?," *Fortean Times,* 1989, p. 48.

141 These accusations, if proven: The fullest account of the charges can be found in F. Hoyle and N. C. Wickramasinghe, *Archaeopteryx: A Case of Fossil Forgery,* 1986. The charges appear in: R. S. Watkins, F. Hoyle, N. C. Wickramasinghe, J. Watkins, R. Rabilizirov, and L. M. Spetner, "*Archaeopteryx*—a photographic study," *Brit. J. Photog.,* 1985; R. S. Watkins, F. Hoyle, N. C. Wickramasinghe, J. Watkins, R. Rabilizirov, and L. M. Spetner, "*Archaeopteryx*—a further commentary," *Brit. J. Photog.,* 1985; R. S. Watkins, F. Hoyle, N. C. Wickramasinghe, J. Watkins, R. Rabilizirov, and L. M. Spetner, "*Archaeopteryx*—further evidence," *Brit. J. Photog.,* 1985; R. S. Watkins, F. Hoyle, N. C. Wickramasinghe, J. Watkins, R. Rabilizirov, and L. M. Spetner, "*Archaeopteryx*—problems and a motive," *Brit. J. Photog.,* 1985; F. Hoyle, N. C. Wickramasinghe, and

R. Rabilizirov, "*Archaeopteryx is* a forgery," *Brit. J. Photog.*, 1987; L. M. Spetner, F. Hoyle, N. C. Wickramasinghe, and M. Margaritz, "*Archaeopteryx*— more evidence for a forgery," *Brit. J. Photog.*, 1988. Replies to the charges can be found in: M. E. Howgate, "*Archaeopteryx* counterview," *Brit. J. Photog.*, 1985; A. Charig, J. F. Greenway, A. C. Milner, C. A. Walker, and P. J. Whybrow, "*Archaeopteryx* is not a forgery," *Science*, 1986; S. Rietschel, "False forgery," in M. K. Hecht, et al., eds., *The Beginnings of Birds,* 1985.

141 "looked under magnification like flattened blobs . . .": R. W. Watkins, et al., "Photographic study," 1985, p. 265.

142 "reptilian fossils . . . subsequently reclassified . . . ," "remain unique in that . . .": Ibid., pp. 264–65.

142 "too small . . . ," "foreign material," "This reply also shows . . .": L. M. Spetner, et al., "More evidence," 1988, pp. 15–16.

143 It is standard protocol: A technical guide to SEM work on fossils is J. J. Rose, "A replication technique for scanning electron microscopy: applications for anthropologists," *Amer. J. Phys. Anthrop.*, 1983.

143 "We therefore think . . .": L. M. Spetner, et al., "More evidence," 1988, pp. 15–16.

144 new photographs were published: S. Rietschel, "False forgery," 1985.

145 Siegfried Rietschel: Ibid.

145 Such hairline cracks: See A. Charig, et al., "Not a forgery," 1986, for a description and illustration of these cracks.

147 In 1979, Alan Feduccia and Harrison Tordoff: A. Feduccia and H. Tordoff, "Feathers of *Archaeopteryx*: Asymmetric vanes indicate aerodynamic function," *Science*, 1979.

148 "The paper was actually rejected . . .": A. Feduccia, interview with author.

148 The Eichstätt specimen: P. Wellnhofer, "Eine neues Exemplar von *Archaeopteryx*," *Archaeopteryx*, 1988.

148 Also, the Solnhofer Aktien-Verein specimen: P. Wellnhofer, "A new specimen of *Archaeopteryx* from the Solnhofen limestone," in K. E. Campbell, Jr., ed., *Papers on Avian Paleontology Honoring Pierce Brodkorb*, Science Series, Nat. Hist. Mus. L.A. County, 1992; P. Wellnhofer, "Das siebte Exemplar von *Archaeopteryx* aus den Solnhofener Schichten," *Archaeopteryx*, 1993.

149 "The shape and general proportions . . .": A. Feduccia and H. Tordoff, "Feathers," 1979, p. 1022.

149 "Their aerodynamic design is . . .": A. Feduccia, "On why the dinosaur lacked feathers," in M. K. Hecht, et al., eds., *The Beginnings of Birds*, 1985, p. 78.

149 In these and other works, Feduccia: A. Feduccia, *The Age of Birds*, 1980; A. Feduccia, *The Origin and Evolution of Birds*, 1996.

149 feathers-as-flight-mechanisms thesis: K. C. Parkes, "Speculations on the origin of feathers," *Living Bird*, 1966.

149 furculas in dinosaurs: D. Chure and J. Madsen, "On the presence of furculae in some non-maniraptoran theropods," *J. Vert. Paleo.*, 1996; H. N. Bryant and A. P. Russell, "The occurrence of clavicles within Dinosauria: implications for the homology of

the avian furcula and the utility of negative evidence," *J. Vert. Paleo.*, 1993.

150 Writing with Storrs Olson: S. Olson and A. Feduccia, "Flight capability and the pectoral girdle of *Archaeopteryx*," *Nature*, 1979.

151 Feduccia's contention: J. R. Speakman and S. C. Thomson, "Flight capabilities of *Archaeopteryx*," *Nature*, 1994.

152 "You have asymmetry going . . .": A. Feduccia, interview with author, 10/26/96.

152 "Everything about feathers . . .": A. Feduccia, interview with author, 10/20/96.

153 Heilmann, Savile, and others: G. Heilmann, *Origin of Birds*, 1926; D. B. O. Savile, "Adaptive evolution in the avian wing," *Evol.*, 1957; D. B. O. Savile, "Gliding and flight in the vertebrates," *Am. Zool.*, 1962.

154 "Why not hair?": This question is attributed to K. C. Parkes, "Speculations," 1966, by P. J. Regal, "Common sense and reconstructions of the biology of fossils: *Archaeopteryx* and feathers," in M. K. Hecht, et al., eds., *The Beginnings of Birds*, 1985, p. 72, but Parkes does not use this precise phrase; see also P. J. Regal, "The evolutionary origin of feathers," *Quart. Rev. Biol.*, 1975.

154 "His main claim is . . .": P. J. Regal, "Common sense," 1985, pp. 71, 73.

155 "Birds are basically flying animals . . .": P. J. Regal, "Evolutionary origin," 1975, p. 38.

155 "This line of reasoning . . .": Ibid., p. 38, italics in original.

156 Regal was led to wonder: Ibid., pp. 36ff.

157 In the wild, basking lizards: P. J. Regal, "Common sense," 1985, p. 72.

158 both endothermic and ectothermic animals: Ibid., p. 73; see also P. J. Regal, "Evolutionary origin," 1975.

158 In 1958, an inventive researcher: R. B. Cowles, "Possible origin of dermal temperature regulation," *Evol.*, 1958.

Chapter 7. On the Wing

160 "What good is half an eye . . .": J. Kingsolver and M. Koehl, "Aerodynamics, thermoregulation, and the evolution of insect wings: different scaling and evolutionary change," *Evol.*, 1985, p. 425.

161 Mimi Koehl is a nonconformist: The following discussion is drawn from J. Kingsolver and M. Koehl, "Selective factors in the evolution of insect wings," *Ann. Rev. Entomol.*, 1994, and J. Kingsolver and M. Koehl, "Aerodynamics," 1985.

161 "There is no insect *Archaeopteryx*": The original remark is attributed to R. J. Wootton and C. P. Ellington, "Biomechanics and the origin of insect flight," in J. M. V. Rayner and R. J. Wootton, eds., *Biomechanics in Evolution*, 1991, pp. 99–112, and is repeated in J. Kingsolver and M. Koehl, "Selective factors," 1994, p. 429.

162 "bounded ignorance," "Rather than attempt . . .": J. Kingsolver and M. Koehl, "Selective factors," 1994, p. 426.

162 Wings are used in sexual displays: Ibid., p. 433.

164 A similar idea about theropod evolution: G. S. Paul, *Predatory Dinosaurs of the World: A Complete Illustrated Guide*, 1988, p. 44.

164 "Another paper suggested . . .": M. Koehl, interview with author.

165 Some researchers were so surprised: C. P. Ellington, "Aerodynamics and the origin of insect flight," *Adv. Insect Physiol.,* 1991.

166 "[T]his scenario suggests that . . .": J. Kingsolver and M. Koehl, "Aerodynamics," 1985, p. 502.

167 "One cannot predict . . .": Ibid., pp. 503–504.

170 "We have observed thousands . . .": J. Marden and M. G. Kramer, "Surface-skimming stoneflies: A possible intermediate stage in insect flight evolution," *Science,* 1994, p. 427.

Chapter 8. One Fell Swoop

176 Nonetheless, in a survey: A. Schultz, "Notes on diseases and healed fractures of wild apes and their bearing on the antiquity of pathological conditions in man," *Bull. Hist. of Med.,* 1939, p. 576.

176 The idea of an arboreal origin: G. Heilmann, *Origin of Birds,* 1926.

178 In the case of the giant red flying squirrel: K. D. Scholey, "The climbing and gliding locomotion of the giant red flying squirrel *Petaurista petaurista,*" in W. G. Nachtigall, ed., *Biona Report 4: Fledermausflug,* 1985; K. D. Scholey, "The evolution of flight in bats," in W. G. Nachtigall, ed., *Biona Report 4: Fledermausflug,* 1985.

179 "Climbing certainly has to be . . .": D. S. Pieters, "Functional and constructive limitations in the early evolution of birds," in M. K. Hecht, et al., eds., *The Beginnings of Birds,* 1985, p. 247.

180 Using radio telemetry: P. Jouvertin and H. Weimerskich, "Satellite tracking of wandering albatrosses," *Nature,* 1990.

180 Smaller birds also practice: U. M. Norberg, *Vertebrate Flight,* 1990, pp. 157ff.

180 Whatever the bird's size: C. Pennycuick, "The mechanics of bird migration," *Ibis,* 1969; C. Pennycuick, "Thermal soaring compared in three dissimilar tropical bird species, *Fregata magnificens, Pelecanus occidentalis,* and *Coragyps atratus,*" *J. exp. Biol.,* 1983.

181 Technically, gliding involves: J. A. Oliver, "'Gliding' in amphibians and reptiles, with a remark on an arboreal adaptation in the lizard, *Anolis carolinensis carolinensis* Voigt," *Am. Nat.,* 1951; J. M. V. Rayner, "Flight adaptations in vertebrates," *Symp. Zool. Soc. Lond.,* 1981.

181 Following a seminal paper: W. Bock, "The role of adaptive mechanisms in the origin of higher levels of organization," *Syst. Zool.,* 1965; W. Bock, "The origin and radiation of birds," *Ann. N.Y. Acad. Sci.,* 1969; W. Bock, "The arboreal origin of avian flight," in K. Padian, ed., *Origin of Birds,* 1986.

181 Parachuting is defined: J. A. Oliver, "'Gliding' in amphibians and reptiles, with a remark on an arboreal adaptation in the lizard, *Anolis carolinensis carolinensis* Voigt," *Am. Nat.,* 1951; J. M. V. Rayner, "Flight adaptations in vertebrates," *Symp. Zool. Soc. Lond.,* 1981.

181 Although house cats are not generally: J. Diamond, "Why cats have nine lives," *Nature,* 1988.

182 Several factors determine: J. M. V. Rayner, "Flight adaptations in vertebrates," *Symp. Zool. Soc. Lond.,* 1981; R. Balda, et al., "Gliding to flapping sequence," 1985.

182 "intuitively attractive," "Gliding allows gravity . . .": J. M. V. Rayner, "Mechanical and ecological constraints on flight evolution," in M. K. Hecht, et al., eds., *The Beginnings of Birds,* 1985, pp. 281, 283.

183 Indeed, Bock used his paper: W. Bock, "Adaptive mechanisms," *Syst. Zool.,* 1965.

183 "The one thing they all have . . . ," "almost nothing": A. Feduccia, interview with author, October 20, 1996.

184 "If you can imagine . . .": L. D. Martin, interview with author, 12/1/95.

184 "Walter's method is to . . . ," "The other problem arises . . .": K. Padian, interview with author, 12/3/96.

185 Padian and Rayner: K. Padian, "A functional analysis of flying and walking in pterosaurs," *Paleobiol.,* 1983; J. M. V. Rayner, "Form and function in avian flight," *Curr. Ornithol.,* 1988; J. M. V. Rayner, "Flight adaptations," 1981.

185 Though both gliders and flappers: R. W. Thorington, Jr., and L. R. Heaney, "Body proportions and gliding adaptations of flying squirrels (Petauristinae)," *J. Mammal.,* 1981.

186 For example, measurements of ten flightless birds: B. Livezey, "Flightlessness in the Galapagos cormorant (*Compsohalieus [Nannopterum] harrisi*): heterochrony, giantism, and specialization," *Zool. J. Linn. Soc.,* 1992.

187 "in any transition . . .": R. Balda, et al., "Gliding to flapping sequence," 1985, pp. 275ff.

188 Without dispute, bats evolved: J. M. V. Rayner, "Mechanical constraints," 1985; J. M. V. Rayner, "Vertebrate flapping, flight mechanics and the evolution of flight in bats," in W. G. Nachtigall, ed., *Biona Report 4: Fledermausflug,* 1985; K. D. Scholey, "Flight in bats," 1985; K. Padian, "A comparative phylogenetic and functional approach to the origin of vertebrate flight," in M. B. Fenton, P. Racey, and J. M. V. Rayner, eds., *Recent Advances in the Study of Bats,* 1987.

188 Several scientists have made an intensive study: This information is summarized in G. Viohl, "Geology of the Solnhofen lithographic limestone and the habitat of *Archaeopteryx,*" in M. K. Hecht, et al., eds., *The Beginnings of Birds,* 1985; P. de Buisonjé, "Climatological conditions during deposition of the Solnhofen limestones," in M. K. Hecht, et al., eds., *The Beginnings of Birds,* 1985; K. W. Barthel, N. H. M. Swinburne, and S. Conway Morris, *Solnhofen: A Study in Mesozoic Paleontology,* 1990.

188 Few life forms: P. de Buisonjé, "Climatological conditions," 1985.

189 Such specimens are the less complete: G. Viohl, "Solnhofen lithographic limestone," 1985.

190 "cannot be used to prove . . .": A. Feduccia, *Origin and Evolution,* 1996, p. 109.

190 In support of this scenario: K. W. Barthel, et al., *Solnhofen,* 1990, p. 86.

191 rudimentary glider, good glider: R. Balda, et al., "Gliding to flapping sequence," 1985, p. 269.

191 While this is an effective strategy: Ibid., p. 270.

191 "As one who much prefers . . .": D. Yalden, "Forelimb function in *Archaeopteryx*," in M. K. Hecht, et al., eds., *The Beginnings of Birds,* 1985, pp. 91–92.

193 These show neither blunting nor wear: J. M. V. Rayner, "Avian flight evolution and the problem of *Archaeopteryx*," in J. M. V. Rayner and R. J. Wootton, eds., *Biomechanics in Evolution,* 1991, p. 191.

193 A few years after Yalden: A. Feduccia, "Claw geometry," 1993.

194 Feduccia agrees with Charles Sibley: C. Sibley and J. E. Ahlquist, *Phylogeny and Classification of Birds,* 1990; A. Feduccia, *Age of Birds,* 1980, pp. 162ff.

195 "Most likely, *Archaeopteryx* used . . .": A. Feduccia, "Claw geometry," 1993, p. 792.

196 Comparison of the hoatzin's DNA: S. B. Hedges, M. D. Simmons, M. A. M. van Dijk, G.-J. Caspers, W. W. deJong, and C. Sibley, "Phylogenetic relationships of the hoatzin, an enigmatic South American bird," *Proc. Natl. Acad. Sci.,* 1995.

197 After the hatchling phase: A. Feduccia, *Origin and Evolution,* 1996.

197 The similarity and rarity of clawed wings: P. Brodkorb, "Origin and evolution of birds," in D. S. Farmer and J. R. King, eds., *Avian Biology,* 1971.

197 The risk of damaging the primary feathers: J. Gauthier and K. Padian, "Phylogenetic," 1985, p. 194.

198 To fit between the extended wings: D. Yalden, "What size was *Archaeopteryx?*," *Zool. J. Linn. Soc.,* 1984, p. 180.

198 Happily, the plant fossils: G. Viohl, "Solnhofen lithographic limestone," 1985.

198 The most ingenious proposal: S. Rietschel, "Feathers and wings," 1985.

Chapter 9. Dragons Fly

202 In 1879, Samuel Williston: S. Williston, "Are birds derived from dinosaurs?," *Kansas City Rev. Sci.,* 1879; see also B. F. Mudge, "Are birds derived from dinosaurs?," *Kansas City Rev. Sci.,* 1879.

202 Nopsca's great contribution: F. Nopsca, "Ideas," 1907; see also F. Nopsca, "On the origin of flight in birds," *Proc. Zool. Soc. Lond.,* 1923.

203 *"from the mechanical standpoint . . . ,"* "A patagium is a soft flexible . . .": F. Nopsca, "Ideas," 1907, p. 223.

204 "If we . . . now suppose . . .": Ibid., pp. 234–36.

205 Certainly feathers and reptilian scales: P. J. Regal, "Common sense," 1985.

207 "Perhaps Nopsca was misled . . .": J. Ostrom, "Origin of flight," 1974, p. 31.

207 Ostrom's updated scenario: Ostrom has written many papers explaining his ideas. Among the most important are: "Origin of flight," 1974; "Origin of birds," 1976; "Bird flight: how did it begin?," *Amer. Sci.,* 1979; "Meaning," 1985; "Question," 1991; "Avian flight," 1994.

208 *"Deinonychus* is about my size": L. D. Martin, *Paleoworld,* broadcast on 3/17/96.

208 Bipedality was *obligate:* J. Ostrom, "Question," 1991.

210 modeling by Caple, Balda, and Willis: G. Caple, et al., "Leaping animals," 1983.

210 However, Jacques Gauthier of Yale University: J. Gauthier, "Saurischian mono-

phyly and the origin of birds," in K. Padian, ed., *The Origin of Birds and the Evolution of Flight*, 1986; J. Gauthier and K. Padian, "Phylogenetic," 1985.

210 "the benchmark to which . . . ," "Those who disagree . . .": L. Witmer, "Perspectives," 1991, pp. 427–66.

212 *Archaeopteryx bavarica:* P. Wellnhofer, "Das siebte Exemplar," 1993.

212 incapable of producing a wing flip: R. Vasquez, "Functional osteology," 1992.

212 it may exist in incipient form: K. P. Dial, "Nonsteady flight," 1992.

214 "Characters may aid in recognition . . .": J. Gauthier, "Saurischian monophyly," 1986, p. 36.

214 "It has been estimated . . .": C. McGowan, *Dinosaurs, Spitfires and Sea Dragons*, 1991, pp. 135–36.

214 Reptiles are often cited: A good review of homeothermy, endothermy, and ectothermy for a popular audience can be found in J. R. Horner and J. Gorman, *Digging*, 1988, pp. 174–84.

215 Galapagos tortoises: R. S. Mackay, "Galapagos tortoise and marine iguana deep body temperatures measured by radio telemetry," *Nature*, 1964.

215 No single piece of evidence: J. Horner, interview with author.

216 they cannot maintain a high level: J. Ruben, "Reptilian physiology and the flight capacity of *Archaeopteryx*," *Evol.*, 1991.

217 Finally, de Beer found the bony crest: G. de Beer, *Archaeopteryx lithographica*, 1954.

217 notably Derik Yalden and Alan Feduccia and Storrs Olson: D. Yalden, "The flying ability of *Archaeopteryx*," *Ibis*, 1971; S. Olson and A. Feduccia, "Pectoral girdle," 1979.

Chapter 10. Pathways to the Skies

220 What were pterosaurs like?: A good summary for the layperson can be found in C. Zimmer, "Masters of an ancient sky," *Discover*, 1994.

221 Pterosaurs, birds, and bats: Excellent discussions can be found in many places, including: U. M. Norberg, *Vertebrate Flight*, 1990; K. Padian, "The flight of the pterosaurs," *Nat. Hist.*, 1988; K. Padian and J. M. V. Rayner, "The wings of pterosaurs," *Amer. J. Sci.*, 1993; K. Padian, "Flying and walking in pterosaurs," *Paleobiol.*, 1983; K. Padian, "The origins and aerodynamics of flight in extinct vertebrates," *Palaeont.*, 1985; J. M. V. Rayner, "Flight adaptations," 1981; J. M. V. Rayner, "The evolution of vertebrate flight," *Biol. J. Linn. Soc.*, 1988.

223 "sustained flapping flight . . .": J. Ruben, "The evolution of endothermy in mammals and birds: From physiology to fossils," *Ann. Rev. Physiol.*, 1995, p. 75; see also A. F. Bennett and J. Ruben, "Endothermy and activity in vertebrates," *Science*, 1979.

223 "[T]heir structure is such . . .": E. H. Hankin and D. M. S. Watson, "On the flight of pterodactyls," *The Aeronautical Journal*, 1914, p. 324.

223 They were envisioned: G. A. Hazlehurst and J. M. V. Rayner, "Flight character-

istics of Jurassic and Triassic Pterosauria: an appraisal based on wing shape," *Paleobiol.*, 1992; R. M. Alexander, "The flight of the pterosaur," *Nature*, 1994.

223 The best-preserved specimen: K. A. von Zittel, "Über Flugsaurier aus dem lithographischen Schiefer Bayerns," *Paläontographica*, 1882.

224 "All flying animals . . .": K. Padian, "The wings of pterosaurs: A new look," *Discovery*, 1979, p. 26.

225 Padian found that Peter Wellnhofer: P. Wellnhofer, "Die Pterodactyloidea (Pterosauria) der Oberjura-Plattenkalke Suddeutschlands," *Abhandlungen der Bayerischen Akademie der Wissenschaften zu München, Mathematisch-Naturwissenschaftlichen Klasse*, 1970.

225 If the bat-like wing: See the excellent article by K. Padian, "The case of the bat-winged pterosaur," in S. J. Czerkas and E. C. Olson, eds., *Dinosaurs Past and Present*, 1987.

226 In 1836, William Buckland: W. Buckland, *Geology and Mineralogy, Considered with Reference to Natural Theology. Bridgewater Treatises on the Power, Wisdom, and Goodness of God as Manifested in the Creation*, vol. 5, 1836.

226 "as aids in sustaining . . .": R. Owen, *A Monograph on the Fossil Reptilia of the Liassic Formations*, 1870.

227 "The starting point . . .": K. Padian, "Bat-winged pterosaur," 1987, pp. 77–78.

228 He had already rejected: K. Padian, interview with author.

228 These hindlimb structures: K. Padian, "Flight of pterosaurs," 1988.

228 This brought Padian to: See K. Padian, "Origins and aerodynamics," 1985, pp. 415ff. for this argument and discussion. Additional information can be found in K. Padian, "Flight of pterosaurs," 1988; K. Padian, "Bat-winged pterosaur," 1987; K. Padian, "Pterosaurs: were they functional birds or functional bats?," in J. M. V. Rayner and R. J. Wootton, eds., *Biomechanics in Evolution*, 1991.

230 Since a good supply of oxygen: J. Ruben, "Evolution of endothermy," 1995, p. 75.

230 Pterosaur bones show the cellular structure: A. J. de Ricqlès, K. Padian, and J. Horner, "Paleohistology of pterosaur bones," *J. Vert. Paleo.*, 1993.

230 Also, like bats and birds: F. Broili, "Haare bei Reptilien," *Anatomischer Anzeiger*, 1941; P. Wellnhofer, *The Illustrated Encyclopedia of Pterosaurs*, 1991.

231 Not everyone accepts: D. M. Unwin and N. Bakhurina, "*Sordes pilosus* and the nature of the pterosaur flight apparatus," *Nature*, 1994; see also R. M. Alexander, "Flight of pterosaur," *Nature*, 1994.

231 I was once told a sad story: A. C. Walker, pers. comm. to author.

231 "Sunscreen . . .": J. Horner, interview with author.

231 "[In these specimens] the main wing membrane . . .": D. M. Unwin and N. Bakhurina, "*Sordes pilosus*," 1994, pp. 62–64.

231 "I find myself . . .": D. Unwin, quoted in C. Zimmer, "Masters," 1994, p. 51.

232 What Padian proposes: K. Padian and J. M. V. Rayner, "The wings of pterosaurs," *Amer. J. Sci.*, 1993; P. Wellnhofer, "Die Pterodactyloidea," 1970.

233 Another challenge: J. Ruben and W. Hillenius, "Pterosaurs as ectotherms," *J. Vert. Paleo.*, 1996; J. Ruben, "Evolution of endothermy," 1995; J. Ruben, W.

Hillenius, N. Geist, A. Leitch, T. Jones, P. Currie, J. Horner, and G. Espe III, "The metabolic status of some Late Cretaceous dinosaurs," *Science*, 1996. For popular accounts, see M. W. Browne, "Flying reptiles pose evolutionary puzzle," *The New York Times*, 1996, and J. Fischman, "Were dinos cold-blooded after all? The nose knows," *Science*, 1995.

234 "What you have to have . . .": J. Horner, interview with author.

234 "Rosetta Stone": J. Ruben, quoted in J. Fischman, "Were dinos cold-blooded?," 1995.

234 "can be reliable . . . ," "extremely variable . . .": G. Paul, "The status of respiratory turbinates in theropods," *J. Vert. Paleo.*, 1996.

235 Pennycuick uses the giant red: C. Pennycuick, "Mechanical constraints," 1986.

235 A radical variant: D. Schaller, "Wing evolution," in M. K. Hecht, et al., *The Beginnings of Birds,* 1985.

237 Hanging is so important: W. A. Schutt, Jr., "Digital morphology in the Chiroptera: The passive digital lock," *Acta Anatomica,* 1993.

237 The reason is that bats: C. Pennycuick, "Mechanical constraints," 1986.

237 Still, bats are more restricted: U. M. Norberg and J. M. V. Rayner, "Ecological morphology in bats (Mammalia: Chiroptera): wing adaptations, flight performance, foraging strategy and echolocation," *Phil. Trans. Roy. Soc. Lond.*, 1987; J. M. V. Rayner, "The mechanics of flapping flight in bats," in M. B. Fenton, P. A. Racey, and J. M. V. Rayner, eds., *Recent Advances in the Study of Bats*, 1987; J. M. V. Rayner, "Evolution of vertebrate flight," *Biol. J. Linn. Soc.*, 1988.

237 It seems likely that: P. Pirlot, "Wing design and the origin of bats," in M. K. Hecht, P. Goody, and B. M. Hecht, eds., *Major Patterns in Vertebrate Evolution,* 1977.

238 pelicans, condors, and albatrosses: C. Pennycuick, "Power requirements for horizontal flight in pigeons," *J. exp. Biol.,* 1968; U. M. Norberg, *Vertebrate Flight,* 1990, p. 178.

238 The extinct New World teratorns: K. E. Campbell and E. P. Tonni, "Size and locomotion in teratorns (Aves: Teratornithidae)," *Auk,* 1984.

238 *Quetzalcoatlus northropi* from west Texas: W. Langston, Jr., "Pterosaurs," *Sci. Amer.,* 1981.

239 However, flying at night: C. Pennycuick, "Mechanical constraints," 1986, pp. 84ff.

239 Although the aerodynamic consequence: U. M. Norberg, *Vertebrate Flight,* 1990, pp. 232ff.

239 Most bats have aspect ratios: P. Pirlot, "Wing design," 1977, p. 384.

239 the range in birds: U. M. Norberg, *Vertebrate Flight,* 1990, p. 239.

239 The difference is that such birds: J. M. V. Rayner, "Evolution of vertebrate flight," 1988, pp. 41ff.; P. Pirlot, "Wing design," 1977.

240 *Pteranodon* was the subject: C. Bramwell and G. R. Whitfield, "Biomechanics of *Pteranodon*," *Phil. Trans. Roy. Soc. Lond.*, 1974.

240 In 1983, James Brower: J. Brower, "The aerodynamics of *Pteranodon* and *Nyctosaurus,* two large pterosaurs from the Upper Cretaceous of Kansas," *J. Vert. Paleo.*, 1983.

240 With its low stalling speed: C. McGowan, *Dinosaurs*, 1991, p. 285.

240 Similarly, when Marden reviewed: J. Marden, "From damselflies to pterosaurs: How burst and sustainable flight performance scale with size," *Am. J. Physiol.*, 1994.

241 The contrary is more likely: P. Pirlot, "Wing design," 1977.

241 uncluttered habitats: R. M. Alexander, "Flight of pterosaur," 1994.

Chapter 11. Flying High

247 "A nonflying animal . . .": C. Pennycuick, "Mechanical constraints," 1986, pp. 86–87.

248 The highest reported wing loading: F. A. Humphrey and B. Livezey, "Flightlessness in steamer ducks (Anatidae: *Tachyeres*): Its morphological bases and probable evolution," *Evol.*, 1985.

248 Another important predictor: The following discussion is taken from C. Pennycuick, "Mechanical constraints," 1986.

249 Jeremy Rayner speculates that: J. M. V. Rayner, "Mechanical constraints," 1985.

250 The weight of *Archaeopteryx:* H. Jerison, "Brain evolution and *Archaeopteryx*," *Nature*, 1968.

250 "obviously avian," "from reconstructions": Ibid., p. 1381.

250 The body weight of *Archaeopteryx:* D. Yalden, "Flying ability," *Ibis*, 1971; D. Yalden, "Flying ability of *Archaeopteryx*," *Nature*, 1971; D. Yalden, "What size," 1984.

251 Because Crawford Greenewalt: C. Greenewalt, "Dimensional relationships for flying animals," *Smithson. Misc. Collns.*, 1962. This is the source for the equations used to predict wing area, wing loading, aspect ratio, and wingspan for different groups of birds.

253 "*Archaeopteryx* is envisaged as . . .": R. Thulborn and T. Hamley, "A new paleoecological role for *Archaeopteryx*," in M. K. Hecht, et al., eds., *The Beginnings of Birds*, 1985, p. 81.

254 Stalling speed is a function: K. Padian, "Origins and aerodynamics," 1985, see pp. 425ff.

254 Since a small, bipedal: C. R. Taylor, "Energy costs of animal locomotion," in L. Botis, ed., *Comparative Physiology*, 1973.

255 Since 1954, when Gavin de Beer: G. de Beer, *Archaeopteryx lithographica*, 1954.

255 A similar conclusion: S. Olson and A. Feduccia, "Pectoral girdle," 1979.

255 However, Ostrom counterargued: J. Ostrom, "Hypothetical anatomical stages," *Smithson. Contrib. to Paleobiol.*, 1976.

256 "*Archaeopteryx* was an active . . . ," "[T]he concept of flight . . .": P. Dodson, "Conference report: International *Archaeopteryx* conference," *J. Vert. Paleo.*, 1985, p. 179.

257 Rietschel's main paper: S. Rietschel, "Feathers and wings," 1985, see especially

Fig. 4; see also P. Shipman, "*Archaeopteryx*'s wings and flight capability," in preparation.

259 The Norbergs: U. M. Norberg, *Vertebrate Flight,* 1990.

259 "ground foraging . . .": Ibid., pp. 241–43, for this and other quotations in this and the subsequent paragraph.

Chapter 12. The Tangled Wing

262 Marden took an empirical approach: J. Marden, "Maximum lift production during takeoff in flying animals," *J. exp. Biol.,* 1987.

264 For example, Pennycuick: C. Pennycuick, "Bird migration," 1969.

264 "These birds should be . . .": J. Marden, "Maximum lift," 1987, p. 248.

265 "Consequently [argues Ruben] . . .": J. Ruben, "Reptilian physiology," p. 6.

265 "Certainly, compared to modern birds . . .": Ibid., pp. 5–6, italics in original.

266 Ruben arbitrarily set, "a little less than half": J. Ruben, interview with author.

268 As always when the subject: J. R. Speakman, "Flight capabilities of *Archaeopteryx,*" *Evol.,* 1993; J. Ruben, "Powered flight in *Archaeopteryx*: Response to Speakman," *Evol.,* 1993.

268 "Larger animals generate . . .": J. Marden, "Maximum lift," 1987, p. 249.

269 With Ruben's estimate: J. Marden, "From damselflies," 1994; see pp. R1079ff. for equations and discussion.

270 In his calculations, Marden used: The 550-pound (250-kilogram) body weight estimate for *Quetzalcoatlus* comes from G. S. Paul, "A reevaluation of the mass and flight of giant pterosaurs," 1990, and G. S. Paul, "The many myths, some old, some new, of dinosaurology," 1991. The estimate of 280 pounds (127 kilograms) was given to me in a personal communication from Wann Langston, Jr., and is more generally accepted than Paul's estimate.

271 "With an animal of entirely anomalous . . .": J. Marden, interview with author.

272 Pterosaurs are built like: T. Lehman and W. Langston, Jr., "Habitat and behavior of *Quetzalcoatlus*: Paleoenvironmental reconstruction of the Javelina Formation (Upper Cretaceous), Big Bend National Park, Texas," *J. Vert. Paleo.,* 1996.

274 "In normal, sane times . . . ," "a sort of halo," "I was forewarned . . .": P. Currie, interview with author, October 22–23, 1996.

276 "The reader can be sure . . .": J. Ostrom, "The osteology of *Compsognathus longipes* Wagner," *Zitteliana,* 1978, pp. 115–16.

276 In his recent book: A. Feduccia, *Origin and Evolution,* 1996, p. 131.

276 "The problem with the thing . . .": A. Feduccia, interview with author.

277 And a new theropod dinosaur: F. E. Novas and P. F. Puerta, "New evidence concerning avian origins from the Late Cretaceous of Patagonia," *Nature,* 1997.

Bibliography

—————

Alexander, R. M. "The flight of the pterosaur." *Nature* 371:12–13, 1994.

Altangerel, P., M. A. Norell, L. M. Chiappe, and J. M. Clark. "Flightless bird from the Cretaceous of Mongolia." *Nature* 362:623–26, 1993.

Anonymous. Press release, American Museum of Natural History, December 20, 1995.

Bakker, R. T. "The superiority of dinosaurs." *Discovery* 3(1):11–22, 1968.

————. "Dinosaur physiology and the origin of mammals." *Evol.* 25:636–58, 1971.

————. "Anatomical and ecological evidence of endothermy in dinosaurs." *Nature* 238:81–85, 1972.

Balda, R., G. Caple, and W. Willis. "Comparison of the gliding to flapping sequence with the flapping to gliding sequence." In M. K. Hecht, J. Ostrom, G. Viohl, and P. Wellnhofer, eds., *The Beginnings of Birds*. Eichstätt: Freunde des Jura-Museums, 1985.

Barber, L. *The Heyday of Natural History: 1820–1970*. Garden City: Doubleday, 1980.

Barthel, K. W., N. H. M. Swinburne, and S. C. Morris. *Solnhofen: A Study in Mesozoic Palaeontology*. Cambridge: Cambridge University Press, 1990. Translation of Barthel, K. W. *Solnhofen*. Thun, Switzerland: Ott Verlag + Druck AG, 1978.

Beeton, I. *The Book of Household Management*. London: Ward, Lock and Tyler, 1861.

Bennett, A. F., and J. Ruben. "Endothermy and activity in vertebrates." *Science* 206:649–54, 1979.

Benton, M. "Origin and interrelationships of dinosaurs." In D. B. Weishampel, P.

Dodson, and H. Osmolska, eds., *The Dinosauria*. Berkeley: University of California Press, 1990.

Blixen, K. *Out of Africa*. London: Jonathan Cape, 1937; reprinted 1973.

Bock, W. "The role of adaptive mechanisms in the origin of higher levels of organization." *Syst. Zool.* 14:272–87, 1965.

———. "The origin and radiation of birds." *Ann. N.Y. Acad. Sci.* 167:147–55, 1969.

———. "The arboreal origin of avian flight." In K. Padian, ed., *The Origin of Birds and the Evolution of Flight*. San Francisco: Calif. Acad. Sci., 1986.

———, and W. D. Miller. "The scansorial foot of the woodpeckers with comments on the evolution of perching and climbing feet in birds." *Am. Mus. Nat. Hist. Novitates* 1931:1–45, 1959.

Boggs, D. F., and K. P. Dial. "Neuromuscular organization and regional EMG activity of the pectoralis in the pigeon." *J. Morphol.* 218:43–57, 1993.

Bramwell, C., and G. R. Whitfield. "Biomechanics of *Pteranodon*." *Phil. Trans. Roy. Soc. Lond.* B267:503–81, 1974.

Brodkorb, P. "Origin and evolution of birds." In D. S. Farmer and J. R. King, eds., *Avian Biology* I. New York: Academic Press, 1971.

Broili, F. "Haare bei Reptilien." *Anatomischer Anzeige* 92:62–68, 1941.

Broom, R. "On the early development of the appendicular skeleton of the ostrich, with remarks on the origin of birds." *Trans. S. Afr. Phil. Soc.* 16:355–68, 1906.

———. "On the South African pseudosuchian *Euparkeria* and allied genera." *Proc. Zool. Soc. Lond.* 619–33, 1913.

Brower, J. "The aerodynamics of *Pteranodon* and *Nyctosaurus*, two large pterosaurs from the Upper Cretaceous of Kansas." *J. Vert. Paleo.* 3:84–124, 1983.

Browne, M. W. "Flying reptiles pose evolutionary puzzle." *The New York Times*, October 22, 1996, C1.

Bryant, H. N., and A. P. Russell. "The occurrence of clavicles within Dinosauria: implications for the homology of the avian furcula and the utility of negative evidence." *J. Vert. Paleo.* 13:171–84, 1993.

Buckland, W. *Geology and Mineralogy, Considered with Reference to Natural Theology. Bridgewater Treatises on the Power, Wisdom, and Goodness of God as Manifested in the Creation.* Vol. 5. London, 1836.

Burkhard, S. "Remarks on reconstruction of *Archaeopteryx* wing." In M. K. Hecht, J. Ostrom, G. Viohl, and P. Wellnhofer, eds., *The Beginnings of Birds*. Eichstätt: Freunde des Jura-Museums, 1985.

Campbell, K. E., and E. P. Tonni. "Size and locomotion in teratorns (Aves: Teratornithidae)." *Auk* 100:390–403, 1984.

Caple, G., R. Balda, and W. Willis. "The physics of leaping animals and the evolution of preflight." *Amer. Nat.* 121:455–67, 1983.

Charig, A., J. F. Greenway, A. C. Milner, C. A. Walker, and P. J. Whybrow. "*Archaeopteryx* is not a forgery." *Science* 232:622–25, 1986.

Chatterjee, S. "Cranial anatomy and relationships of a new Triassic bird from Texas." *Phil. Trans. Roy. Soc. Lond.* B332:277–346, 1991.

Chiappe, L. "The first 85 million years of avian evolution." *Nature* 378:349–55, 1995.

Colbert, E., and M. Morales. *Evolution of the Vertebrates.* New York: John Wiley & Sons, Inc., 1991.

Chure, D., and J. Madsen. "On the presence of furculae in some non-maniraptoran theropods." *J. Vert. Paleo.* 16(3):573–77, 1996.

Cope, E. D. "An account of the extinct reptiles which approached the birds." *Proc. Acad. Natl. Sci. Philadelphia* 19:234–35, 1867.

Coues, E. "On the mechanism of flexion and extension in birds' wings." *Proc. Amer. Assn. Adv. Sci.* 20:278–84, 1871.

Cowles, R. B. "Possible origin of dermal temperature regulation." *Evol.* 12:347–57, 1958.

Darwin, C. *The Origin of Species by Means of Natural Selection or the Preservation of Favored Races in the Struggle for Life.* London: 1859; reprinted New York: New American Library, 1958.

Darwin, F., ed. *Charles Darwin: His Life Told in an Autobiographical Chapter and in a Selected Series of His Published Letters.* New York: Appleton, 1892. Reprinted as F. Darwin, ed., *The Autobiography of Charles Darwin and Selected Letters.* New York: Dover Publications, 1958.

de Beer, Gavin. *Archaeopteryx lithographica.* London: British Museum (Natural History), 1954.

de Buisonjé, P. "Climatological conditions during deposition of the Solnhofen limestones." In M. K. Hecht, J. Ostrom, G. Viohl, and P. Wellnhofer, eds., *The Beginnings of Birds.* Eichstätt: Freunde des Jura-Museums, 1985.

de Ricqlès, A. J. "Evolution of endothermy: Histological evidence." *Evolutionary Theory* 1:51–80, 1974.

———, K. Padian, and J. Horner. "Paleohistology of pterosaur bones." *J. Vert. Paleo.* 13, Suppl. to No. 3: 54A, 1993.

Desmond, A. *Huxley: The Devil's Disciple.* London: Michael Joseph, 1994.

———, and J. Moore. *Darwin.* New York: Warner Books, 1991.

di Gregorio, M. *T. H. Huxley's Place in Natural Science.* New Haven: Yale University Press, 1984.

Dial, K. P. "Activity patterns of the wing muscles of the pigeon (*Columbia livia*) during different modes of flight." *J. exp. Zool.* 262:357–73, 1992.

———. "Avian forelimb muscles and nonsteady flight: Can birds fly without using the muscles in their wings?" *Auk* 109:874–85, 1992.

Diamond, J. "Why cats have nine lives." *Nature* 332:586–88, 1988.

Dodson, P. "Conference report: International *Archaeopteryx* conference." *J. Vert. Paleo.* 5:177–79, 1985.

Ellington, C. P. "Aerodynamics and the origin of insect flight." *Adv. Insect Physiol.* 23:171–210, 1991.

Feduccia, A. *The Age of Birds.* Cambridge: Harvard University Press, 1980.

———. "On why the dinosaur lacked feathers." In M. K. Hecht, J. Ostrom, G.

Viohl, and P. Wellnhofer, eds., *The Beginnings of Birds.* Eichstätt: Freunde des Jura-Museums, 1985.

———. "Evidence from claw geometry indicating arboreal habits of *Archaeopteryx.*" *Science* 259:790–93, 1993.

———. "The great dinosaur debate." *Living Bird* 13:28–33, 1994.

———. "Explosive evolution in Tertiary birds and mammals." *Science* 267:637–38, 1995.

———. *The Origin and Evolution of Birds.* New Haven: Yale University Press, 1996.

———, and H. Tordoff. "Feathers of *Archaeopteryx*: Asymmetric vanes indicate aerodynamic function." *Science* 203:1021–22, 1979.

Fischman, J. "Were dinos cold-blooded after all? The nose knows." *Science* 270:735–36, 1995.

Fisher, H. L. "Bony mechanisms of automatic flexion and extension in the pigeon's wing." *Science* 126:446, 1957.

Gatesy, S. M. "Caudofemoral musculature and the evolution of theropod locomotion." *Paleobiol.* 16(2):170–86, 1990.

———, and K. P. Dial. "Tail muscle activity patterns in walking and flying pigeons (*Columbia livia*)." *J. exp. Biol.* 176:55–76, 1993.

———. "Locomotor modules and the evolution of avian flight." *Evol.* 50:331–40, 1996.

Gauthier, J. "Saurischian monophyly and the origin of birds." In Kevin Padian, ed., *The Origin of Birds and the Evolution of Flight.* San Francisco: California Acad. of Sci., 1–55, 1986.

———. "A cladistic analysis of the higher systematic categories of the Diapsida." Ph.D. thesis, University of California, Berkeley, 1994.

———, and K. Padian. "Phylogenetic, functional, and aerodynamic analyses of the origin of birds and their flight." In M. K. Hecht, J. Ostrom, G. Viohl, and P. Wellnhofer, eds., *The Beginnings of Birds.* Eichstätt: Freunde des Jura-Museums, 1985.

Goldschmidt, O. *An History of the Earth and Animated Nature.* London, 1779.

Gordon, J. E. *Structures: or Why Things Don't Fall Down.* Harmondsworth, Middlesex: Penguin Books Ltd., 1978.

Goslow, G. E., Jr., K. P. Dial, and F. A. Jenkins, Jr. "Bird flight: Insights and complications." *BioScience* 40(2):108–15, 1990.

Greenewalt, C. "Dimensional relationships for flying animals." *Smithson. Misc. Collns.* 144(2):1–46, 1962.

Hankin, E. H., and D. M. S. Watson. "On the flight of pterodactyls." *The Aeronautical Journal* 18:324–35, 1914.

Hart, L. B. *The World of Leonardo da Vinci: Man of Science, Engineer and Dreamer of Flight.* New York: Viking Press, 1961.

Hartmann, F. A. "Locomotor mechanisms of birds." *Smithson. Misc. Colln.* 143:1–91, 1961.

Hazlehurst, G. A., and J. M. V. Rayner. "Flight characteristics of Jurassic and Triassic Pterosauria: an appraisal based on wing shape." *Paleobiol.* 18:447–63, 1992.

Hecht, M. K., and S. Tarsitano. "The paleobiology and phylogenetic position of *Archaeopteryx.*" *Geobios Mem. Sp.* 6:141–49, 1982.

———. "Paleontological myopia." *Nature* 309:588, 1984.

Hedenström, Å. "Migration by soaring or flapping flight in birds: the relative importance of energy cost and speed." *Phil. Trans. Roy. Soc. Lond.* B342:353–61, 1993.

Hedges, S. B., M. D. Simmons, M. A. M. van Dijk, G.-J. Caspers, W. W. deJong, and C. Sibley. "Phylogenetic relationships of the hoatzin, an enigmatic South American bird." *Proc. Natl. Acad. Sci.* 92:11662–65, 1995.

Heilmann, G. *Origin of Birds.* London: Witherby, 1926.

Herzog, K. *Anatomie und Flugbiologie der Vögel.* Stuttgart: Gustav Fischer, 1968.

Hildebrand, M. *Analysis of Vertebrate Structure.* New York: John Wiley & Sons, Inc., 1974.

Hinchcliffe, J. R. "'One, two, three' or 'Two, three, four': An embryologist's view of the homologies of the digits and carpus of modern birds." In M. K. Hecht, J. Ostrom, G. Viohl, and P. Wellnhofer, eds., *The Beginnings of Birds.* Eichstätt: Freunde des Jura-Museums, 1985.

———, and Hecht, M. K. "Homology of the bird wing skeleton: embryological versus paleontological evidence." *Evol. Biol.* 18:21–39, 1984.

Horner, J. R., and J. Gorman. *Digging Dinosaurs.* New York: Workman Publishing, 1988.

Horner, J. R., and R. Makela. "Nest of juveniles provides evidence of family structure among dinosaurs." *Nature* 282:296–98, 1979.

Howard, F. *Wilbur and Orville: A Biography of the Wright Brothers.* New York: Ballantine Books, 1987.

Howgate, M. E. "*Archaeopteryx*—no new finds after all." *Nature* 306:644, 1983.

———. "*Archaeopteryx*'s morphology." *Nature* 310:104, 1984.

———. "*Archaeopteryx* counterview." *Brit. J. Photog.* 132:348, 29 March 1985.

Hoyle, F., and N. C. Wickramasinghe. *Archaeopteryx: A Case of Fossil Forgery.* Swansea: Christopher Davis, 1986.

Hoyle, F., N. C. Wickramasinghe, and R. Rabilizirov. "*Archaeopteryx is* a forgery." *Brit. J. Photog.* 134:682, 12 June 1987.

Humphrey, F. A., and B. Livezey. "Flightlessness in steamer ducks (Anatidae: *Tachyeres*): Its morphological bases and probable evolution." *Evol.* 40:540–58, 1985.

Huxley, T. H. "On the animals which are most nearly intermediate between the dinosaurian reptiles and birds." *Sci. Mem.* 3, 1868.

———. "Remarks upon *Archaeopteryx lithographica.*" *Proc. Roy. Soc. Lond.* 15:243–48, 1868.

———. "Further evidence of the affinity between the dinosaurian reptiles and birds." *Quart. J. Geol. Soc. Lond.* 26:12–31, 1870.

Jenkins, F. A., Jr., K. P. Dial, and G. E. Goslow, Jr. "A cineradiographic analysis of bird flight: The wishbone in starlings is a spring." *Science* 241:1495–98, 1988.

Jerison, H. "Brain evolution and *Archaeopteryx.*" *Nature* 219:1381–82, 1968.

Bibliography

Jouvertin, P., and H. Weimerskich. "Satellite tracking of wandering albatrosses." *Nature* 343:746–48, 1990.

Kandon, K. *Vertebrates*. Dubuque, Ia.: Wm. C. Brown Communications, Inc., 1995.

Kingsolver, J., and M. Koehl. "Aerodynamics, thermoregulation, and the evolution of insect wings: different scaling and evolutionary change." *Evol.* 39:488–504, 1985.

———. "Selective factors in the evolution of insect wings." *Ann. Rev. Entomol.* 39:425–51, 1994.

Langston, W., Jr. "Pterosaurs." *Sci. Amer.* 244:122–36, 1981.

Lehman, T., and W. Langston, Jr. "Habitat and behavior of *Quetzalcoatlus*: Paleoenvironmental reconstruction of the Javelina Formation (Upper Cretaceous), Big Bend National Park, Texas," *J. Vert. Paleo.* 16(3):48A, 1996.

Leonardo da Vinci. Ms. E., fol. 54r and *Codex Atlanticus* 45r.

Livezey, B. "Flightlessness in the Galapagos cormorant (*Compsohalieus [Nannopterum] harrisi*): heterochrony, giantism, and specialization." *Zool. J. Linn. Soc.* 105:155–224, 1992.

Lucas, A. M., and P. Stettenheim. *Avian Anatomy: Integument*. 2 vols. Agricultural Handbook #362. Washington, D.C.: U.S. Government Printing Office, 1972.

Mackay, R. S. "Galapagos tortoise and marine iguana deep body temperatures measured by radio telemetry." *Nature* 204:355–58, 1964.

Marden, J. "Maximum lift production during takeoff in flying animals." *J. exp. Biol.* 130:235–58, 1987.

———. "From damselflies to pterosaurs: How burst and sustainable flight performance scale with size." *Am. J. Physiol.* 266:R1077–84, 1994.

———, and M. G. Kramer. "Surface-skimming stoneflies: A possible intermediate stage in insect flight evolution." *Science* 266:427–30, 1994.

Martin, L. D. "The origin and early radiation of birds." In A. H. Brush and G. A. Clark, Jr., eds., *Perspectives in Ornithology*. Cambridge: Cambridge University Press, 1983.

———. "The origin of birds and early flight." *Curr. Ornithol.* 1:105–29, 1983.

———. "The relationship of *Archaeopteryx* to other birds." In M. K. Hecht, J. Ostrom, G. Viohl, and P. Wellnhofer, eds., *The Beginnings of Birds*. Eichstätt: Freunde des Jura-Museums, 1985.

———. "The beginning of the modern avian radiation." *L'Évolution des Oiseaux d'après le témoignage des fossiles*. In C. Mourer-Chauviré, ed., *Docum. Lab. Géol. Lyon*, 99:9–19. Lyon: Université Claude Bernard, 1987.

———. "Mesozoic birds and the origin of birds." In H.-P. Schultze and L. Trueb, eds., *Origins of the Higher Groups of Tetrapods*. Ithaca: Comstock Press, 1991.

———, and J. D. Stewart. "Homologies in the avian tarsus." *Nature* 315:519, 1985.

———, and K. Whetstone. "The origin of birds: structure of the tarsus and teeth." *Auk* 97:86–93, 1980.

Mayr, E. *The Growth of Biological Thought: Diversity, Evolution and Inheritance*. Cambridge, Mass.: Belknap Press, 1982.

McGowan, C. "Evolutionary relationships of ratites and carinates: Evidence from ontogeny of the tarsus." *Nature* 307:733–35, 1984.

———. "Homologies in the avian tarsus, McGowan replies." *Nature* 315:159–60, 1985.

———. *Dinosaurs, Spitfires and Sea Dragons*. Cambridge: Harvard University Press, 1991.

Mitchell, J. "When feathers fly: A case of fossil forgery?" *Fortean Times*, Summer 1989, 47–51.

Morell, V. "*Archaeopteryx*: Early bird catches a can of worms." *Science* 259:764–65, 1993.

Mudge, B. F. "Are birds derived from dinosaurs?" *Kansas City Rev. Sci.* 3:224–26, 1879.

Nopsca, F. "Ideas on the origin of flight." *Proc. Zool. Soc. Lond.* (15):223–36, 1907.

———. "On the origin of flight in birds." *Proc. Zool. Soc. Lond.* 463–77, 1923.

Norberg, R. Å. "Function of vane asymmetry and shaft curvature in bird flight feathers; inferences on flight ability of *Archaeopteryx*." In M. K. Hecht, J. Ostrom, G. Viohl, and P. Wellnhofer, eds., *The Beginnings of Birds*. Eichstätt: Freunde des Jura-Museums, 1985.

Norberg, U. M. *Vertebrate Flight*. Berlin: Springer-Verlag, 1990.

Norberg, U. M., and J. M. V. Rayner. "Ecological morphology in bats (Mammalia: Chiroptera): wing adaptations, flight performance, foraging strategy and echolocation." *Phil. Trans. Roy. Soc. Lond.* B316:335–427, 1987.

Norell, M. A., J. M. Clark, D. Demberelyin, B. Rhinchen, L. M. Chiappe, A. R. Davidson, M. McKenna, P. Altangerel, and M. J. Novacek. "A theropod dinosaur embryo and the affinities of the Flaming Cliffs dinosaur eggs." *Science* 266:779–82, 1994.

Norell, M. A., J. M. Clark, R. Weintraub, L. M. Chiappe, and D. Demberelyin. "A nesting dinosaur." *Nature* 378:774–76, 1995.

Novas, F. E., and P. F. Puerta. "New evidence concerning avian origins from the Late Cretaceous of Patagonia." *Nature* 387:390–92, 1997.

Oliver, J. A. "'Gliding' in amphibians and reptiles, with a remark on an arboreal adaptation in the lizard, *Anolis carolinensis carolinensis* Voigt." *Am. Nat.* 85:171–76, 1951.

Olson, S., and A. Feduccia. "Flight capability and the pectoral girdle of *Archaeopteryx*." *Nature* 278:247–48, 1979.

Ostrom, J. "Terrible claw." *Discovery* 5(1):1–9, 1969.

———. "Terrestrial vertebrates as indicator of Mesozoic climates." *N. Amer. Paleo. Conv. Proc. (Chicago, 1969):* 347–76, 1970.

———. "The ancestry of birds." *Nature* 242:136, 1973.

———. "*Archaeopteryx* and the origin of flight." *Quart. Rev. Biol.* 49:27–47, 1974.

———. "Some hypothetical anatomical stages in the evolution of avian flight." Collected Papers in Avian Paleontology Honoring the 90th Birthday of Alexander Wetmore, S. L. Olson, ed., *Smithson. Contrib. to Paleobiol.* 27:1–21, 1976.

———. "*Archaeopteryx* and the origin of birds." *Biol. J. Linnean Soc.* 8:91–182, 1976.

————. "The osteology of *Compsognathus longipes* Wagner." *Zitteliana* 4:73–118, 1978.

————. "Bird flight: How did it begin?" *Amer. Sci.* 67:46–56, 1979.

————. "Introduction to *Archaeopteryx*." In M. K. Hecht, J. Ostrom, G. Viohl, and P. Wellnhofer, eds., *The Beginnings of Birds*. Eichstätt: Freunde des Jura-Museums, 1985.

————. "The meaning of *Archaeopteryx*." In M. K. Hecht, J. Ostrom, G. Viohl, and P. Wellnhofer, eds., *The Beginnings of Birds*. Eichstätt: Freunde des Jura-Museums, 1985.

————. "The Yale *Archaeopteryx*: The one that flew the coop." In M. K. Hecht, J. Ostrom, G. Viohl, and P. Wellnhofer, eds., *The Beginnings of Birds*. Eichstätt: Freunde des Jura-Museums, 1985.

————. "The question of the origin of birds." In H.-P. Schultze and L. Trueb, eds., *Origins of the Higher Groups of Tetrapods*. Ithaca: Comstock Press, 1991.

————. "On the origin of birds and of avian flight." In R. S. Spencer, ed., *Major Features of Vertebrate Evolution*. Short Courses in Paleontology 7:160–77, 1994.

Owen, R. "On the *Archaeopteryx* of von Meyer, with a description of the fossil remains of a long-tailed species, from the lithographic stone of Solnhofen." *Phil. Trans. Roy. Soc. Lond.* 153:33–47, 1863.

————. *A Monograph on the Fossil Reptilia of the Liassic Formations*, Pt. 3. London: Palaeontographical Society, 1870.

Padian, K. "The wings of pterosaurs: A new look." *Discovery* 14(1):21–29, 1979.

————. "Running, leaping, lifting off." *The Sciences,* May/June 1982, 10–15.

————. "A functional analysis of flying and walking in pterosaurs." *Paleobiol.* 9(3):218–39, 1983.

————. "The origins and aerodynamics of flight in extinct vertebrates." *Palaeont.* 28(3):413–33, 1985.

————. "A comparative phylogenetic and functional approach to the origin of vertebrate flight." In M. B. Fenton, P. Racey, and J. M. V. Rayner, eds., *Recent Advances in the Study of Bats*. Cambridge: Cambridge University Press, 1987.

————. "The case of the bat-winged pterosaur." In S. J. Czerkas and E. C. Olson, eds., *Dinosaurs Past and Present*, vol. II. Los Angeles: Natural History Museum of Los Angeles, 1987.

————. "The flight of the pterosaurs." *Nat. Hist.* 12/88:58–65, 1988.

————. "Pterosaurs: were they functional birds or functional bats?" In J. M. V. Rayner and R. J. Wootton, eds., *Biomechanics in Evolution*. Cambridge: Cambridge University Press, 145–60, 1991.

————, and J. M. V. Rayner. "The wings of pterosaurs." *Amer. J. Sci.* 293A:91–166, 1993.

Parkes, K. C. "Speculations on the origin of feathers." *Living Bird* 5:77–86, 1966.

Paul, G. S. *Predatory Dinosaurs of the World: A Complete Illustrated Guide*. New York: New York Acad. Sci., 1988.

————. "A reevaluation of the mass and flight of giant pterosaurs." *J. Vert. Paleo.* 10, suppl. to no. 3:37, 1990.

———. "The many myths, some old, some new, of dinosaurology." *Mod. Geol.* 16:69–99, 1991.

———. "The status of respiratory turbinates in theropods." *J. Vert. Paleo.* 16, suppl. to No. 3:57A, 1996.

Pennycuick, C. "Power requirements for horizontal flight in pigeons." *J. exp. Biol.* 49:527–55, 1968.

———. "The mechanics of bird migration." *Ibis* 111:525–56, 1969.

———. "Soaring behavior and performance of some East African birds, observed from a motor-glider." *Ibis* 114:178–218, 1972.

———. "Thermal soaring compared in three dissimilar tropical bird species, *Fregata magnificens, Pelecanus occidentalis,* and *Coragyps atratus.*" *J. exp. Biol.* 102:307–25, 1983.

———. "Mechanical constraints on the evolution of flight." In K. Padian, ed., *The Origin of Birds and the Evolution of Flight.* San Francisco: Calif. Acad. Sci., 1986.

———. *Bird flight performance: A practical calculation manual.* Oxford: Oxford Univ., 1989.

Pieters, D. S. "Functional and constructive limitations in the early evolution of birds." In M. K. Hecht, J. Ostrom, G. Viohl, and P. Wellnhofer, eds., *The Beginnings of Birds.* Eichstätt: Freunde des Jura-Museums, 1985.

Pirlot, P. "Wing design and the origin of bats." In M. K. Hecht, P. Goody, and B. M. Hecht, eds., *Major Patterns in Vertebrate Evolution.* New York: Plenum, NATO ASI 14:375–426, 1977.

Rayner, J. M. V. "Flight adaptations in vertebrates." *Symp. Zool. Soc. Lond.* 48:137–172, 1981.

———. "Mechanical and ecological constraints on flight evolution." In M. K. Hecht, J. Ostrom, G. Viohl, and P. Wellnhofer, eds., *The Beginnings of Birds.* Eichstätt: Freunde des Jura-Museums, 1985.

———. "Vertebrate flapping, flight mechanics and the evolution of flight in bats." In W. G. Nachtigall, ed., *Biona Report 4: Fledermausflug.* Stuttgart: Gustav Fischer, 1985.

———. "The mechanics of flapping flight in bats." In M. B. Fenton, P. A. Racey, and J. M. V. Rayner, eds., *Recent Advances in the Study of Bats.* Cambridge: Cambridge University Press, 1987.

———. "Form and function in avian flight." *Curr. Ornithol.* 5:1–66, 1988.

———. "Avian flight evolution and the problem of Archaeopteryx." In J. M. V. Rayner and R. J. Wootton, eds., *Biomechanics in Evolution.* Cambridge: Cambridge University Press, 1991.

Regal, P. J. "The evolutionary origin of feathers." *Quart. Rev. Biol.* 50(1):35–66, 1975.

———. "Common sense and reconstructions of the biology of fossils: *Archaeopteryx* and feathers." In M. K. Hecht, J. Ostrom, G. Viohl, and P. Wellnhofer, eds., *The Beginnings of Birds.* Eichstätt: Freunde des Jura-Museums, 1985.

Rietschel, S. "False forgery." In M. K. Hecht, J. Ostrom, G. Viohl, and P. Wellnhofer, eds., *The Beginnings of Birds.* Eichstätt: Freunde des Jura-Museums, 1985.

————. "Feathers and wings of *Archaeopteryx,* and the question of her flight ability." In M. K. Hecht, J. Ostrom, G. Viohl, and P. Wellnhofer, eds., *The Beginnings of Birds.* Eichstätt: Freunde des Jura-Museums, 1985.

Rose, J. J. "A replication technique for scanning electron microscopy: applications for anthropologists." *Amer. J. Phys. Anthrop.* 62:255–61, 1983.

Ruben, J. "Reptilian physiology and the flight capacity of *Archaeopteryx.*" *Evol.* 45:1–17, 1991.

————. "Powered flight in *Archaeopteryx*: Response to Speakman." *Evol.* 47:935–38, 1993.

————. "The evolution of endothermy in mammals and birds: From physiology to fossils." *Ann. Rev. Physiol.* 57:69–95, 1995.

————, and W. Hillenius. "Pterosaurs as ectotherms." *J. Vert. Paleo.* 16, suppl. to #3:61A, 1996.

Sanz, J. L., L. M. Chiappe, B. P. Pérez-Moreno, J. J. Moratalla, F. Hernández-Carrasquilla, A. D. Buscalioni, F. Ortega, F. J. Poyato-Ariza, D. Russkin-Gutman, and X. Martínez-Delclòs, "A nesting bird from the Lower Cretaceous of Spain," *Science* 276:1543–46, 1997.

Savile, D. B. O. "Adaptive evolution in the avian wing." *Evol.* 11:212–24, 1957.

————. "Gliding and flight in the vertebrates." *Am. Zool.* 2:161–66, 1962.

Schaller, D. "Wing evolution." In M. K. Hecht, J. Ostrom, G. Viohl, and P. Wellnhofer, eds., *The Beginnings of Birds.* Eichstätt: Freunde des Jura-Museums, 1985.

Scholey, K. D. "The climbing and gliding locomotion of the giant red flying squirrel *Petaurista petaurista.*" In W. G. Nachtigall, ed., *Biona Report 4: Fledermausflug.* Stuttgart: Gustav Fischer, 1985.

————. "The evolution of flight in bats." In W. G. Nachtigall, ed., *Biona Report 4: Fledermausflug.* Stuttgart: Gustav Fischer, 1985.

Schultz, A. "Notes on diseases and healed fractures of wild apes and their bearing on the antiquity of pathological conditions in man." *Bull. Hist. of Med.* 7:571–82, 1939.

Schutt, W. A., Jr. "Digital morphology in the Chiroptera: The passive digital lock." *Acta Anatomica* 148:219–27, 1993.

Shipman, P., A. Walker, and D. Bichell. *The Human Skeleton.* Cambridge, Mass.: Harvard University Press, 1985.

Sibley, C., and J. E. Ahlquist. *Phylogeny and Classification of Birds.* New Haven: Yale University Press, 1990.

Speakman, J. R. "Flight capabilities of *Archaeopteryx.*" *Evol.* 47:336–40, 1993.

————, and S. C. Thomson. "Flight capabilities of *Archaeopteryx.*" *Nature* 370:524, 1994.

Spetner, L. M., F. Hoyle, N. C. Wickramasinghe, and M. Margaritz. "Archaeopteryx—more evidence for a forgery." *Brit. J. Photog.* 135:14–17, 7 Jan 1988.

Stearn, W. T. *The Natural History Museum at South Kensington: A History of the British Museum (Natural History) 1753–1980.* London: Heinemann, 1981.

Sy, M.-H. "Funktionell-anatomische Untersuchungen am Vögelflugel." *J. Orn., Lpz.* 84:199–296, 1936.

Tarsitano, S. "The morphological and aerodynamic constraints on the origin of avian flight." In M. K. Hecht, J. Ostrom, G. Viohl, and P. Wellnhofer, eds., *The Beginnings of Birds*. Eichstätt: Freunde des Jura-Museums, 1985.

———. "*Archaeopteryx*—Quo Vadis?" In H.-P. Schultze and Linda Trueb, eds., *The Origins of the Higher Groups of Tetrapods*. Ithaca: Comstock Press, 1991.

———, and M. K. Hecht. "A reconsideration of the reptilian relationships of *Archaeopteryx*." *Zool. J. Linn Soc.* 69:149–82, 1980.

Taylor, C. R. "Energy costs of animal locomotion." In L. Botis, ed. *Comparative Physiology*. Amsterdam: North Holland, 1973.

Thorington, R. W., Jr., and L. R. Heaney. "Body proportions and gliding adaptations of flying squirrels (Petauristinae)." *J. Mammal.* 62:101–14, 1981.

Thulborn, R. A., and T. L. Hamley. "The reptilian relationships of *Archaeopteryx*." *Aust. J. Zool.* 30:611–34, 1982.

———. "A new paleoecological role for *Archaeopteryx*." In M. K. Hecht, J. Ostrom, G. Viohl, and P. Wellnhofer, eds., *The Beginnings of Birds*. Eichstätt: Freunde des Jura-Museums, 1985.

Unwin, D. M., and N. Bakhurina. "*Sordes pilosus* and the nature of the pterosaur flight apparatus." *Nature* 371:62–64, 1994.

U.S. Geological Survey. *Geologic Time—The Age of the Earth*. Washington, D.C.: U.S. Geological Survey, 1970.

Vasquez, R. "Functional osteology of the avian wrist and the evolution of flapping flight." *J. Morphol.* 211:259–68, 1992.

Viohl, G. "Geology of the Solnhofen lithographic limestone and the habitat of *Archaeopteryx*." In M. K. Hecht, J. Ostrom, G. Viohl, and P. Wellnhofer, eds., *The Beginnings of Birds*. Eichstätt: Freunde des Jura-Museums, 1985.

———. "Carl F. and Ernst O. Häberlein, the sellers of the London and Berlin specimens of *Archaeopteryx*." In M. K. Hecht, J. Ostrom, G. Viohl, and P. Wellnhofer, eds., *The Beginnings of Birds*. Eichstätt: Freunde des Jura-Museums, 1985.

Wagner, J. A. "Über ein neues, angeblich mit Vögelfedern versehenes Reptil aus dem Solnhofener lithographischen Schiefer." *Sitzungsber. Bayer. Akad. Wiss.* 2:146–54, 1861; W. S. Dallas, transl., "On a new fossil reptile supposed to be furnished with feathers," *Ann. Mag. Nat. Hist.* (3)9:261–67, 1862.

Walker, A. C. "Locomotor adaptations in recent and fossil Madagascan lemurs." Ph.D. thesis, University of London, 1967.

Walker, A. D. "New light on the origin of birds and crocodiles." *Nature* 237:257–63, 1972.

———. "Evolution of the pelvis in birds and dinosaurs." In S. Andrews, R. Miles, and A. D. Walker, eds., *Problems in Vertebrate Evolution*. London: Academic Press, 319–58, 1974.

Watkins, R. S., F. Hoyle, N. C. Wickramasinghe, J. Watkins, R. Rabilizirov, and L. M. Spetner. "*Archaeopteryx*—a photographic study." *Brit. J. Photog.* 132:264–66, 8 March 1985.

———. "*Archaeopteryx*—a further commentary." *Brit. J. Photog.* 132:358–60, 29 March 1985.

———. "*Archaeopteryx*—further evidence." *Brit. J. Photog.* 132:469–71, 26 April 1985.

———. "*Archaeopteryx*—problems and a motive." *Brit. J. Photog.* 132:693–95, 703, 21 June 1985.

Wellnhofer, P. "Die Pterodactyloidea (Pterosauria) der Oberjura-Plattenkalke Suddeutschlands." *Abhandlungen der Bayerischen Akademie der Wissenschaften zu München, Mathematisch-Naturwissenschaftlichen Klasse* 141:1–133, 1970.

———. "Eine neues Exemplar von *Archaeopteryx*." *Archaeopteryx* 6:1–30, 1988.

———. *The Illustrated Encyclopedia of Pterosaurs.* London: Salamander Books, 1991.

———. "A new specimen of *Archaeopteryx* from the Solnhofen limestone." In K. E. Campbell, Jr., ed., *Papers on Avian Paleontology Honoring Pierce Brodkorb,* Science Series Nat. Hist. Mus. L.A. County, 36:3–23, 1992.

———. "Das siebte Exemplar von *Archaeopteryx* aus den Solnhofener Schichten." *Archaeopteryx* 11:21–48, 1993.

Williston, S. "Are birds derived from dinosaurs?" *Kansas City Rev. Sci.* 3:457–60, 1879.

Witmer, L. "Perspectives on avian origins." In H.-P. Schultze and L. Trueb, eds., *Origins of the Higher Groups of Tetrapods.* Ithaca: Comstock Press, 1991.

Wohl, R. *A Passion for Wings: Aviation and the Western Imagination, 1908–1918.* New Haven: Yale University Press, 1994.

Wootton, R. J., and C. P. Ellington. "Biomechanics and the origin of insect flight." In J. M. V. Rayner and R. J. Wootton, eds., *Biomechanics in Evolution.* Cambridge: Cambridge University Press, 99–112, 1991.

Yalden, D. "The flying ability of *Archaeopteryx*." *Ibis* 113:349–56, 1971.

———. "What size was *Archaeopteryx*?" *Zool. J. Linn. Soc.* 82:177–88, 1984.

———. "Forelimb function in *Archaeopteryx*." In M. K. Hecht, J. Ostrom, G. Viohl, and P. Wellnhofer, eds., *The Beginnings of Birds.* Eichstätt: Freunde des Jura-Museums, 1985.

Zimmer, C. "Ruffled feathers." *Discover* 13(5):44–54, 1992.

———. "Masters of an ancient sky." *Discover* 15(2):42–54, 1994.

Zittel, K. A. von. "Über Flugsaurier aus dem lithographischen Schiefer Bayerns." *Paläontographica* 29:47–80, 1882.

Index

Picture Credits

Figure 1. After U.S. Geological Survey, *Geologic Time—The Age of the Earth*, Washington, D.C.: U.S. Geological Survey, 1970.

Figure 2. Copyright Peter Wellnhofer.

Figures 3, 4, 6, 7, 8, 9, 30, 57. Courtesy of John Ostrom.

Figures 5 (left), 39. After A. M. Lucas and P. Stettenheim, *Avian Anatomy: Integument*, 2 vols., Agricultural Handbook #362, Washington, D.C.: U.S. Government Printing Office, 1972, courtesy of P. Stettenheim.

Figure 5 (right). After P. Shipman, A. Walker, and D. Bichell, *The Human Skeleton*, Cambridge, Mass.: Harvard University Press, 1985.

Figures 10, 11A, 58. Reprinted with permission. Photos by F. Höck, copyright Bavarian State Collection.

Figure 11B. Drawing copyright Peter Wellnhofer.

Figure 12. Drawing courtesy A. C. Walker.

Figures 13, 47. Reprinted with permission from U. M. Norberg, *Vertebrate Flight*, 1990, © Springer-Verlag Heidelberg.

Figure 14. Reprinted with permission from K. Kandon, *Vertebrates*, 1995, © Wm. C. Brown Communications, Inc.

Figures 15, 16, 17. Reprinted with permission from F. A. Jenkins, Jr., et al., "A cineradiographic analysis of bird flight: The wishbone in starlings is a spring," *Science* 241:1495–98, 1988. Copyright 1988 American Association for the Advancement of Science.

Figure 18. Reprinted with permission from C. Pennycuick, "Mechanical constraints on the evolution of flight," in K. Padian, ed., *The Origin of Birds and the Evolution of Flight*, 1986, courtesy California Academy of Sciences.

Figure 19. Reprinted with permission from K. P. Dial, "Avian forelimb muscles and nonsteady flight: Can birds fly without muscles in their wings?," *Auk* 109:874–85, 1993.

Figures 20, 26, 29. Reprinted with permission; copyright Greg Paul.

Figure 21. Reprinted with permission from S. M. Gatesy and K. P. Dial, "Tail muscle activity patterns in walking and flying pigeons (*Columbia livia*)," *J. exp. Biol.* 176:55–76, 1993. Published by the Company of Biologists, Ltd.

Figures 22, 23. Reprinted with permission from L. Chiappe, "The first 85 million years of avian evolution," *Nature* 378:349–55, 1995. Copyright 1995 Macmillan Magazines Ltd.

Figures 24, 59, 61, 65, 67. Reprinted with permission; copyright © 1988 D. Braginetz.

Figure 27. After J. Gauthier, "A cladistic analysis of the higher systematic categories of the Diapsida," Ph.D. thesis, University of California, Berkeley, 1984, and J. Gauthier, "Saurischian monophyly and the origin of birds," in K. Padian, ed., *The Origin of Birds and the Evolution of Flight*, 1986. Reprinted with permission from Gatesy, "Caudofemoral musculature and the evolution of theropod locomotion," *Paleobiol.* 16(2):170–86, 1990; courtesy S. Gatesy and editor, *Paleobiology*.

Figure 28. Reprinted with permission from F. Nopsca, "Ideas on the origin of flight," *Proc. Zool. Soc. Lond.* (15):223–36, 1907. Copyright Zoological Society of London.

Figure 31. Reprinted with permission from J. Ostrom, "Bird flight: How did it begin?," *Amer. Sci.* 67:46–56, 1979.

Figures 32, 38. Reprinted with permission from S. Rietschel, "False forgery," in M. K. Hecht et al., eds., *The Beginnings of Birds*, Eichstätt: Freunde des Jura-Museums, 1985.

Figure 33. Reprinted with permission from Colbert and Morales, *Evolution of the Vertebrates*, John Wiley & Sons, Inc., copyright 1991.

Figure 34. Reprinted with permission from F. Nopsca, "On the origin of flight in birds," *Proc. Zool. Soc. Lond.* 463–77, 1923. Copyright Zoological Society of London.

Figures 35, 36, 51. Reprinted with permission from J. Ostrom, "*Archaeopteryx* and the origin of flight," *Quart. Rev. Biol.* 49:27–47, 1974. Copyright 1974 University of Chicago Press.

Figure 37. Reprinted with permission from A. Feduccia, "Evidence from claw geometry indicating arboreal habits of *Archaeopteryx*," *Science* 259:790–93, 1993. Copyright 1993 American Association for the Advancement of Science.

Figure 40. Reprinted with permission from A. Feduccia and H. Tordoff, "Feathers of *Archaeopteryx:* Asymmetric vanes indicate aerodynamic function," *Science* 203:1021–22, 1979. Copyright 1979 American Association for the Advancement of Science.

Figures 41, 42. Reprinted with permission from P. J. Regal, "The evolutionary origin of feathers," *Quart. Rev. Biol.* 50(1):35–66, 1975. Copyright 1975 University of Chicago Press.

Figure 43. Reprinted with permission from J. Kingsolver and M. Koehl, "Selective factors in the evolution of insect wings," *Ann. Rev. Entomol.* 39:425–51, 1994. Copyright Annual Reviews, Inc.

Figures 44, 45, 49. Reprinted with permission from K. W. Barthel, *Solnhofen*, Ott Verlag + Druck AG Thun, 1978, translated as K. W. Barthel et al., *Solnhofen: A Study in Mesozoic Palaeontology*, Cambridge: Cambridge University Press, 1990.

Figure 46. Reprinted with permission from A. C. Walker, "Locomotor adaptations in recent and fossil Madagascan lemurs," Ph.D. thesis, University of London, 1967.

Figure 48. Reprinted from O. Goldschmidt, *An History of the Earth and Animated Nature*, London, 1779.

Figure 50. Reprinted with permission from D. Yalden, "Forelimb function in *Archaeopteryx*," in M. K. Hecht et al., eds., *The Beginnings of Birds*, Eichstätt: Freunde des Jura-Museums, 1985.

Figure 52. Reprinted with permission from A. M. Lucas and P. Stettenheim, *Avian Anatomy: Integument*, 2 vols., Agricultural Handbook #362, Washington, D.C.: U.S. Government Printing Office, 1972; courtesy of P. Stettenheim.

Figure 53. Drawing by G. M. Sutton. Reprinted with permission from Gill, *Ornithology*, 1974, © Academy of Natural Sciences of Philadelphia.

Figure 54. Reprinted with permission from M. Hildebrand, *Analysis of Vertebrate Structure*, 1974, copyright 1974 John Wiley & Sons, Inc.

Figure 56. Reprinted with permission from S. Olson and A. Feduccia, "Flight capability and the pectoral girdle of *Archaeopteryx*," *Nature* 278:247–48, 1979. Copyright 1979 Macmillan Magazines Ltd.

Figures 60, 68. Reprinted with permission; copyright Bavarian State Collection.

Figure 62. Reprinted from S. Th. von Soemmerring, "Über einen *Ornithocephalus brevirostris* der Vorwelt," *Denkschriften der koniglinke Bayerische Akademis der Wissenschaften (math.-phys. Classe)* 6:89–104, 1817.

Figure 63. Reprinted from W. Buckland, *Geology and Mineralogy, Considered with Reference to Natural Theology. Bridgewater Treatises on the Power, Wisdom, and Goodness of God as Manifested in the Creation*, vol. 5, London, 1836.

Figure 64. Reprinted from K. A. von Zittel, "Über Flugsaurier aus dem lithographischen Schiefer Bayerns," *Paläontographica* 29:47–80, 1882.

Figure 66. Reprinted with permission; courtesy of Peabody Museum of Natural History, Yale University.

Figure 69. Photo courtesy of J. Ruben and T. Jones.

Figure 70. Courtesy of Wann Langston, Jr.

Figure 71. Reprinted with permission from U. M. Norberg, *Vertebrate Flight*, 1990, © Springer-Verlag Heidelberg. After K. Herzog, *Anatomie und Flugbiologie der Vögel*, 1968.

Figure 72. Reprinted with permission from D. Yalden, "What size was *Ar-*

chaeopteryx?," *Zool. J. Linn. Soc.* 82:177–88, 1984. Copyright Academic Press, Ltd., London.

Figure 73. Reprinted with permission from R. Thulborn and T. L. Hamley, "A new paleoecological role for *Archaeopteryx*," in M. K. Hecht et al., eds., *The Beginnings of Birds,* Eichstätt: Freunde des Jura-Museums, 1985.

Figure 74. After S. Rietschel, "False forgery," in M. K. Hecht et al., eds., *The Beginnings of Birds,* Eichstätt: Freunde des Jura-Museums, 1985, and D. Yalden, "What size was *Archaeopteryx?*," *Zool. J. Linn. Soc.* 82:177–88, 1984. Copyright Academic Press, Ltd., London. Image courtesy of Vicki Emsch and Richard Sherwood.

Figure 75. Reprinted with permission after U. M. Norberg, *Vertebrate Flight,* 1990, © Springer-Verlag Heidelberg; permission also granted by *J. exp. Biol.* and the Company of Biologists.

Figure 76. Reprinted with permission from J. Marden, "From damselflies to pterosaurs: How burst and sustainable flight performance scale with size," *Am. J. Physiol.* 266:R1077–84, 1994.

Figure 77. Reprinted with permission; copyright 1996 M. W. Skrepnick.